a home

DO BIRDS HAVE KNEES ?

STEPHEN MOSS

B L O O M S B U R Y
LONDON • NEW DELHI • NEW YORK • SYDNEY

giving
nature
a home

The RSPB is the country's largest nature conservation charity, inspiring everyone to give nature a home so that birds and wildlife can thrive again.

By buying this book you are helping to fund the RSPB's conservation work.

If you would like to know more about The RSPB, visit the website at www.rspb.org.uk or write to: The RSPB, The Lodge, Sandy, Bedfordshire, SG19 2DL; 01767 680551.

Bloomsbury Natural History
An imprint of Bloomsbury Publishing Plc

50 Bedford Square
London
WC1B 3DP
UK

1385 Broadway
New York
NY 10018
USA

www.bloomsbury.com

BLOOMSBURY and the Diana logo are trademarks of Bloomsbury Publishing Plc

First published 2016

A catalogue record for this book is available from the British Library.

Library of Congress Cataloguing-in-Publication data has been applied for.

ISBN:	PB:	978-1-4729-3235-8
epub:		978-1-4729-3236-5
ePDF:		978-1-4729-3237-2

2 4 6 8 10 9 7 5 3 1

Design by Rod Teasdale

Printed and Bound in China by RRD Asia Printing Solutions Limited

MIX
Paper from
responsible sources
FSC® C101537

To find out more about our authors and books visit www.bloomsbury.com. Here you will find extracts, author interviews, details of forthcoming events and the option to sign up for our newsletters.

giving
nature
rspb a home

DO BIRDS HAVE KNEES?

STEPHEN MOSS

BLOOMSBURY
LONDON • NEW DELHI • NEW YORK • SYDNEY

giving
nature
a home

The RSPB is the country's largest nature conservation charity, inspiring everyone to give nature a home so that birds and wildlife can thrive again.

By buying this book you are helping to fund the RSPB's conservation work.

If you would like to know more about The RSPB, visit the website at www.rspb.org.uk or write to: The RSPB, The Lodge, Sandy, Bedfordshire, SG19 2DL; 01767 680551.

Bloomsbury Natural History
An imprint of Bloomsbury Publishing Plc

50 Bedford Square	1385 Broadway
London	New York
WC1B 3DP	NY 10018
UK	USA

www.bloomsbury.com

BLOOMSBURY and the Diana logo are trademarks of Bloomsbury Publishing Plc

First published 2016

A catalogue record for this book is available from the British Library.

Library of Congress Cataloguing-in-Publication data has been applied for.

ISBN:	PB:	978-1-4729-3235-8
ePub:		978-1-4729-3236-5
ePDF:		978-1-4729-3237-2

2 4 6 8 10 9 7 5 3 1

Design by Rod Teasdale

Printed and Bound in China by RRD Asia Printing Solutions Limited

MIX
Paper from
responsible sources
FSC
www.fsc.org   FSC® C101537

To find out more about our authors and books visit www.bloomsbury.com. Here you will find extracts, author interviews, details of forthcoming events and the option to sign up for our newsletters.

CONTENTS

What is a bird?   6

Where do birds come from?   42

How many birds are there?   60

Where do birds live?   74

How do birds move?   94

What do birds eat?   116

Why do birds sing?   136

How do birds reproduce?   154

Where do birds go?   188

How do we relate to birds?   206

Bibliography   232

Acknowledgements   233

Picture credits   233

Index   234

INTRODUCTION

Hummingbirds feed on nectar

Questions are a way of focusing our interest, but finding answers can be frustrating. Having spent much of my life fielding other people's questions about birds, I had long yearned for an easy reference source that provided all the answers in a single place. Hence the simple question-and-answer format of this book, which allows any reader to get straight to the answer they want without becoming swamped by information they may not need.

I began the compilation process by collecting 'raw' questions from a broad spectrum of people, including friends and family, beginners and experts, and the staff of my publishers, Bloomsbury. Some were fascinating, some ludicrous and some frankly unanswerable, but all played their part in revealing the kind of things that people would like to know about birds. If you were part of this process, I hope you will find your answer here – and, with luck, a lot more besides.

The questions – more than 500 in total – are arranged in ten chapters, each tackling a major theme, such

as feeding, breeding or migration. A comprehensive index helps you locate any question you want answered. Having done so, I hope you will be drawn further in, finding equally interesting answers on related topics, or simply reading onwards to gain a deeper insight into a particular subject.

During the book's compilation I used a number of reference books in order to finalise questions, check facts and glean ideas. Of these, the most important were those listed in the bibliography by Bird, Brooke & Birkhead, Campbell & Lack, Clements,

Leahy, Todd and Weaver. All published 'facts' were checked against at least two further sources, usually more.

You will find the text liberally sprinkled with headlined boxes containing little nuggets of information. These are what I call 'record breakers', and list superlatives such as the biggest, smallest, highest, fastest and so on. Being records, they are subject to certain qualifications: some, such as those concerned with longevity, may already have been surpassed by the time this book has hit the shelves; others have their absolute accuracy open to question. Many simply reflect what has been measured or studied to date, which means they are by no means the final word on the subject. Facts and statistics – especially those related to the latest scientific discoveries – often show a distinct bias towards European or North American species, simply because this is still where most research takes place. Where there is any measure of doubt, I have couched the information in suitably non-committal terms, such as 'probably', 'it is claimed' etc. If you do discover a newer or more accurate record, please let me know (via the publishers), and I'll be happy to include it in any future editions.

So who exactly is this book for? My long-time friend and birding companion Daniel Osorio gave me a typically backhanded compliment when he said that it would appeal to intelligent, enquiring 11-year-old boys – the same

Emperor Penguins with chick

as he and I were when we first met. I hope that it will also appeal to 11-year-old girls, since there are far too few women birders, and this might just spark an interest that helps redress the balance. Ultimately, however, I would like to think that the book has something to offer to all ages and that it is equally suitable for experienced birders, complete novices and anyone in between.

Most of all, I hope that you enjoy reading it, and that you are motivated to go out into the field and look anew at birds – which, to my mind, are the most elegant, fascinating and delightful of all creatures.

1 · WHAT IS A BIRD?
Physiology

ANATOMY

What is a bird?

A bird is a warm-blooded, egg-laying vertebrate (animal with a backbone). It has a body covered with feathers, and forelimbs modified to form wings. Technically speaking, birds are all members of the class Aves.

All birds – including this Black Kite – have feathers

What makes birds unique?

In a word: feathers. This is the only characteristic unique to birds. Mammals are also warm-blooded, some reptiles and two strange mammals also lay eggs, and bats and some insects can also fly. No other creature has evolved feathers, but all birds – even flightless ones such as Ostriches and penguins – have them.

Macaroni Penguins

Are birds warm-blooded?

Yes, just like mammals, birds are 'warm-blooded' or homeothermic. This means they can (and must) maintain a constant body temperature of between 38°C and 43°C, regardless of the temperature of the air around them. In cold weather, birds help retain this heat by fluffing out their feathers to trap an insulating layer of air next to their warm skin.

Some species also huddle together in communal roosts to avoid losing heat. Birds whose chicks are naked when they hatch (known as 'altricial' or 'nidicolous' species) have to brood their chicks, which remain 'cold-blooded' (poikilothermic) until they fledge. Brooding birds cover their chicks with the soft feathers of their underparts, so their body heat keeps the youngsters warm.

Do birds have a skeleton?

Like all vertebrates, birds have an internal skeleton. However, it has been cunningly customised to suit a bird's unique requirements. Many of the bones are hollow and criss-crossed with internal struts, making them very strong yet incredibly light. This keeps a bird's body weight to a minimum, allowing it to take to the air and fly. A bird's skeleton also has two important modifications: the hind limbs and pelvis have shifted to enable it to walk or hop on two legs; while the forelimbs have been modified into wings, enabling most birds to fly. The huge, keeled breastbone (take a closer look at a roast chicken!) is also a special flight adaptation, since it holds the powerful muscles required for beating the wings.

Chicken skeleton

Big bird

The world's largest living bird is the Ostrich, which can weigh up to 136kg (300lbs) – about 85,000 times as heavy as the world's smallest species (see page 33). One specimen was said to have reached 150kg (330lbs). Unsurprisingly, the Ostrich is also the world's tallest living bird, occasionally reaching a height of 2.5 metres (well over eight feet).

An Ostrich with her brood of chicks

How do birds breathe?

Like mammals, birds have lungs, which extract oxygen from the air, transfer it to the blood, and expel waste carbon dioxide. Unlike mammals, birds also have a secondary system of air sacs located around their body and even inside their bones. This unique adaptation enables birds to circulate oxygen much more efficiently – vital for allowing them to fly without getting out of breath.

Do birds perspire?

No. Birds don't have any sweat glands on their skin, so they lose excess heat by panting or seeking shade.

The heart rate of a hummingbird is faster than that of any other bird

How fast does a bird's heart beat?

This depends upon the bird's size – and what it is doing. Large birds tend to have slow heart rates (the Ostrich's is only about 38 beats per minute), while most songbirds range between 200–500 beats per minute. Hummingbirds may even reach more than 1,000 beats per minute. Heart rates increase in cold weather and when a bird is under stress. Our own hearts average about 72 beats per minute at rest.

Do birds have teeth?

No modern bird has true teeth – these were lost during the evolutionary process of getting light enough for flight. But some species have sharp cutting edges on the mandibles of their bill. A few tropical species such as the African barbets and the unique Tooth-billed Pigeon of Samoa have tooth-like notches on their mandibles, but these are not used for chewing food. A chick ready to hatch has an 'egg-tooth', for chipping its way out. However, this is not a true tooth, and it drops off a few days after hatching.

What is the difference between a 'bill' and a 'beak'?

Except for spelling, nothing at all: the two terms are interchangeable (though birders and ornithologists tend to prefer the term 'bill'). Both refer to the horny projection at the front of every bird's skull, consisting of the upper and lower mandibles – essentially the equivalent of a mammal's jaws.

The Common Snipe has a very long bill in proportion to its body length

9

What is a bird's bill made from?

The bill is bone covered by keratin. This extraordinarily flexible substance can take many forms, enabling bills to tackle all kind of jobs.

Pelicans use the pouch beneath their bill to scoop up food

What do birds use their bills for?

Birds use their bills for catching, carrying and manipulating food, collecting nest material and building nests, excavating nest holes, defending themselves, and preening. Because their forelimbs have been adapted into wings, the tasks we humans perform with our hands and fingers birds must do with their bills. Some birds, such as storks and albatrosses, also clatter their bills together as part of their courtship display.

Why do some birds have such strange-shaped bills?

Every bill has evolved to suit a particular feeding technique, and this has produced some pretty weird shapes. Amongst the strangest belong to Puffins, whose bills have an 'elasticated' base so they can hold several small fish at once; spoonbills, whose spatulate bills contain sensitive nerve endings to detect minute food items; skimmers, whose lower mandible is longer than the upper one, enabling them to 'skim' the surface of the water for morsels of food; and crossbills, whose upper and lower mandibles are crossed to allow them to prise the seeds out of pine cones. But perhaps the weirdest of all is that of a wader found in New Zealand, aptly named the Wrybill. This is the only bird in the world whose bill curves sideways, enabling it to probe for insect food under rocks and stones.

Puffins have large, powerful bills specially designed for holding sand-eels

Roseate Spoonbills, like other members of their family, have a characteristic spatulate bill

Settling the bills

The longest bill belongs to the Australian Pelican, and may reach a length of 48cm (19in). The shortest bill belongs to the Glossy Swiftlet of South-east Asia, at just 4mm (one sixth of an inch) long. The longest bill relative to body size belongs to the Sword-billed Hummingbird of the northern Andes, and measures 9–11cm (3.5–4.3in), more than half the bird's total length. This extraordinary appendage enables the hummingbird to reach nectar hidden deep inside long flowers such as the climbing passion flower. The bill is so heavy that the bird has to hold it at an angle when perched, to avoid toppling over.

The Sword-billed Hummingbird of South America has the longest bill in proportion to its body-length

Why do some birds have deformed bills?

The keratin of a bird's bill grows continually, to compensate for being worn down by use. But if the tip breaks from one mandible, then the tip of the other will have nothing to wear against and so will sometimes grow abnormally long. If a bill becomes twisted to the side then both mandibles can grow, but will cross over. Birds with deformed bills can survive only as long as they are able to feed.

Do birds have tongues?

Yes, they all do, though this organ is more important for some than for others. Most birds do not have much of a sense of taste (see page 27), so many groups, including storks and pelicans, only have very small tongues. Parrots, by contrast, have a large, fleshy tongue to help manipulate their food; hummingbirds have a long, thin one to poke into flowers for nectar; and flamingos have

Tongue-tied

The bird with the longest tongue relative to body size is the Wryneck, a member of the woodpecker family, whose tongue may measure more than 8 cm (over 3 in) – about half its body length.

a whopper, which helps to pump out water while they are filter-feeding. Woodpeckers have proportionally the longest tongue of all; it is rooted at the back of the skull, and its barbed, sticky tip is designed for extracting insects from under loose bark.

Eagles have large, muscular tongues

What do birds use their feet for?

As well as the obvious functions of walking, running, hopping and swimming (see chapter 5), birds use their feet for a number of other purposes. These include perching (all passerines and many other birds), catching food (especially birds of prey and owls), climbing (woodpeckers, parrots and nuthatches) and digging (underground-dwelling species such as the Burrowing Owl). Some species, including various wild game birds and the domestic chicken, even use their feet in combat with rival males – hence the sport of 'cockfighting'.

Male Coots often fight one another, using their powerful feet

How many toes do birds have?

Most birds have four toes on each foot. Some have only three, and the Ostrich has just two. Birds' toes are arranged in one of three configurations: all four pointing forward for gripping onto vertical surfaces (e.g. swifts); three toes pointing forward and one back for perching (passerines and most other birds); or two toes forward and two back for climbing or grasping objects (woodpeckers, cuckoos and parrots, also some Owls). One species, the Osprey, can even adjust the arrangement: normally it points three toes forward and one back, but when

Many waterbirds, like this Black Heron, have long, thin toes, three pointing forward and one back

catching fish it points two forward and two back, to get a better grip on its slippery prey.

What are webbed feet for?

Many unrelated families, including wildfowl (ducks, geese and swans), gulls, cormorants and petrels, have fully webbed feet, which enable them to swim more effectively. Other waterbirds, such as grebes and coots, have developed lobed (partially webbed) feet for the same reason. Webbing is not the only way that feet have adapted to help a bird get around: birds that habitually walk on aquatic vegetation, such as rails and jacanas, have elongated toes to spread

All ducks, like this Mallard, have fully webbed feet

their weight across the surface; while some living in cold climates, such as ptarmigans, have feathered feet which act as snowshoes (and stay warm).

Legging it

The Ostrich has the longest legs of any bird, reaching a massive 120cm (four feet) from hip to toe. The longest legs relative to body size belong to the Black-winged Stilt (and its various relatives), and constitute about 60 per cent of its total length.

Both the Ostrich and the Black-necked Stilt have very long legs

PLUMAGE

Why do birds have feathers?

Feathers do two main jobs for birds: they allow them to fly and they keep them warm. Fossil evidence suggests that feathers probably evolved from reptilian scales, keeping birds' ancestors warm in cold climates. Flight was a much later evolutionary development.

What advantages do feathers bring?

Feathers are miraculously versatile things. Their strength and lightness allow a bird to get airborne without using up too much energy, while their streamlined shape reduces air resistance. They also enable birds to maintain a constant body temperature by dispersing heat in hot weather and trapping heat when it gets chillier. Last, but certainly not least, they are vital for courtship displays, in which one (usually male) bird flaunts his fancy plumage in order to attract a mate and/or repel a rival.

Great Bustards use their feathers in a spectacular courtship display

What are feathers made from?

Feathers are made from a horny substance called keratin: a light, strong and very flexible form of protein. Keratin is also found in a bird's bill and claws – and indeed in human hair and fingernails. Feathers have several component parts: the central shaft, which is hollow at the base and attaches to the bird's skin; the barbs, or side-branches, which are attached to the shaft; and tiny barbules, which branch off the barbs and mesh with each other, giving the feather its unique combination of strength and lightness. Different types of feathers have different uses: large wing feathers enable a bird to fly; contour feathers cover and streamline its body; and soft downy ones keep it warm.

This albatross chick is still covered with down

Bald Eagles have more than 7,000 feathers, more than most birds

How many feathers does a bird have?

This varies enormously, from fewer than 1,000 for some species of hummingbird, to more than 25,000 in the case of the Whistling Swan of North America (most of which are on its head and neck). The number of feathers tends to increase with the size of the bird. So most passerines have between 3,000 and 5,000 feathers, while a Bald Eagle has more than 7,000. Waterbirds tend to have more feathers than landbirds, to help keep them warm and dry.

What are 'primaries', 'secondaries' and 'tertials'?

They are all types of flight feather found in the wing. Primaries are the longest feathers forming the wing tip, secondaries are the shorter ones along the inner part of the wing and tertials are the ones close to the bird's body. Other wing feathers include scapulars and wing coverts.

What is 'bird topography'?

It is simply a technical term for the way we define and name the various external features of a bird's plumage (along with its 'bare parts', such as the bill and legs). Although birds may appear very different in shape and appearance, they all have their feathers arranged in the same basic way. Knowing bird topography helps you navigate around a bird's appearance, and is vital for understanding moult (see page 24). It also helps birders identify similar-looking species by noting subtle differences in their plumage features.

How does a bird look after its feathers?

Feather care is a top priority for any self-respecting bird. Techniques include preening (using the bill to clean individual feathers), scratching (with its feet) and bathing (either in water or using dust). Many birds have a special preen-gland just above the base of their tail. This secretes preen-oil, which they then spread across the surface of their feathers with their bill. It provides waterproofing and may repel parasites.

What happens if a bird damages its feathers?

Birds lead a strenuous life, and feathers take a constant battering; they are often lost or broken, and may become covered with a harmful substance such as tar. Broken or lost feathers

Even waterbirds like this Canada Goose have to bathe

are generally replaced by new ones straightaway. So as long as a bird can still feed and fly it will probably survive, though it may be unable to migrate or get away from predators. However, if a bird's plumage is covered with a contaminant, it may lose its ability to fly or swim, and will be unable to maintain the correct body temperature. Unless it gets clean quickly it will almost certainly die.

How do birds bathe?

Usually, like us, in water. Most birds' favourite method is to partially submerge themselves (either at the edge of a pond, or in a puddle or bird-bath), and splash water over their wings and body while frantically shaking to make sure all the feathers get wet. Many waterbirds, such as gulls and ducks, bathe while sitting on the surface of the water; though some seabirds, such as terns and frigatebirds, are reluctant to do so for fear of becoming waterlogged. Instead, they

plunge down onto the surface of the sea from above, briefly wetting their plumage as they do so. A few species, such as parrots and woodpeckers, deliberately expose themselves to rainfall; while others, including hornbills, rub themselves against wet tropical vegetation.

What about 'dust-bathing'?

Dust-bathing, or 'dusting', is a specialised form of behaviour in which a bird wriggles in dust or sand, tosses the fine particles over its wings and body, rubs them into its plumage, then shakes them out again. This removes excess preen-oil from the feathers and keeps them free of grease. It also helps get rid of harmful parasites such as lice and mites from the feathers and skin. Many species use dusting to keep their plumage in tip-top condition. These include Ostriches, rheas, gamebirds, bustards, sandgrouse, hoopoes and rollers. Fewer passerines do so, but among them are larks, wrens and sparrows.

Tree Sparrow dust-bathing

17

Tails of the unexpected

The bird with the longest feathers of all is the male Crested Argus Pheasant of Southeast Asia. His tail coverts may reach a length of 173cm (68in) – almost three-quarters of his total length. The length increases yearly, with the oldest birds generally having the longest. However, the longest tail feathers ever recorded belonged to an ornamental chicken bred in Japan and were reputed to have reached an amazing 10.5 metres (over 34 feet). The longest tail relative to body size belongs to the male Ribbon-tailed Astrapia, a bird-of-paradise found in New Guinea, which has a tail 90cm (35in) long – three-quarters of its total length. In Europe and North America, the Barn Swallow's tail can grow up to 12.5cm (almost five inches) – two-thirds of its total length.

The Barn Swallow has a very long tail in relation to its body

Great Crested Grebes, like other waterbirds, have a waterproof plumage

Is the plumage of waterbirds waterproof?

In many species, yes. And water does not just flow off a duck's back, but also off the backs of grebes, coots, auks, geese and swans. These birds regularly anoint their plumage with oil from a preen-gland, enabling them to spend long periods on the surface of the water. However, several families that spend much of their time at sea or swimming cannot fully waterproof their plumage. These include frigatebirds, which remain aloft in the air whenever they are away from their nest, and cormorants, which are obliged to stretch out their wings after swimming in order to dry them.

What are the other advantages of well-kept plumage?

A bird's general health owes much to the quality of its plumage, as feathers are vital for many basic life functions such as flight and heat regulation. Male birds with the glossiest, brightest plumage also tend to attract the most females, as healthy plumage is a sign of the bird's ability to reproduce and have lots of healthy offspring.

Do birds carry parasites?

Many birds are crawling with unwanted passengers such as mites, ticks, feather lice and louse flies – parasites that feed on the blood, feathers or shed skin of their host. These generally latch on to a bird while it is still in the nest, and remain there for the rest of its life. Generally they do little harm, unless the bird becomes sick, in which case they may become too numerous for the bird to cope with. Healthy birds

The male Andean Cock-of-the-rock showing off his brightly coloured plumage

keep their parasites in check by regular preening, bathing and dust-bathing. Some, like jays, may also engage in 'anting': a practice of picking up ants and rubbing them into their plumage, or simply allowing ants to swarm all over it. The formic acid secreted by the ants kills many parasites.

Ground shakers

The largest bird that ever lived was probably *Aepyornis Maximus*, or the Elephant Bird of Madagascar. Its weight has been estimated as about 450kg (990lbs) – more than three times as heavy as an Ostrich. It probably survived as late as the 17th century. The giant moas of New Zealand, which became extinct about the same time, were taller, some standing over 3.5 metres (over 11 feet), but weighed a mere 240kg (530lbs).

The Aepyornis, or Elephant Bird

COLOUR

How do birds get their colours?

Some colours are produced by pigments present in feathers; others, such as the iridescent blues, greens and purples of hummingbirds, are created by the refraction of light within the feather structure. Our perception of a bird's colour is also affected by the quality of light: sunrise and sunset make colours appear richer and warmer; while harsh tropical sun may 'burn out' a bird's plumage, making it appear paler than usual.

Why do different birds have different colours?

Because life would be very dull if they all looked the same! Seriously, colour in plumage has several different purposes. Many birds have dull, brownish plumage in order to camouflage themselves against predators, while others sport bright, garish plumage to attract a mate. Sometimes the two can be seen in the same species: the bright colours of a male Mallard contrast dramatically with the subtle brown tones of his mate, who can remain undetected while sitting on her eggs.

Red-spotted Bluethroat

Female Mallard

Blue-and-yellow Macaw

Greater Flamingo

Ptarmigan in spring (top left), summer (top right) and winter (bottom) plumages

How do birds use colours to camouflage themselves?

Some have a 'cryptic' pattern, so they can blend in with their chosen background – as in a nightjar sitting on a branch or a bittern standing among the reeds. But many colourful tropical birds, such as tanagers and parakeets, can also be surprisingly hard to pick out against the pale green foliage of the treetops. Their bold colours and patterns serve to 'disrupt' their outline, making them harder to see. Some birds that spend part of their life in snow adopt an all-white plumage in winter: the Ptarmigan has three different seasonal plumages, each of which enables it to camouflage itself as the landscape changes around it.

Why do so many birds have dark wingtips?

Because melanin, the pigment that makes feathers dark, also makes them stronger. It is an advantage to have stronger feathers at the tips of wings where they are more likely to become worn.

The dark tips to this Common Buzzard's wings help protect the feathers from damage

Why do some birds have bright flashes of colour?

Many birds, especially small passerines such as finches and wheatears, have striking white or yellow patches on their wings, tail or rump. These confuse a predator, by distracting it when the bird takes flight, and also alert others nearby to danger. Often these markings are covered up when the bird is perched, so it can stay hidden from view until it takes off. Bright colours are also often used in breeding displays.

Goldfinches are named after the bright yellow flashes in their wings

What are 'albino', 'melanistic' and 'leucistic' plumages?

These are all abnormal plumage types, caused either by too little or too much of a particular pigment in the feathers. Albinism is caused by a lack of all pigments, making the whole of the bird's plumage appear white and its eyes and other bare parts pink. Melanism is the result of too much of the dark pigment melanin, and makes the bird look darker than usual. Leucism covers variable reduction of plumage pigments, giving the bird a white, piebald or 'washed out' appearance. Xanthochroism refers to an abnormally yellow plumage, and is most often seen in cage birds such as parrots whose diet is lacking in the correct minerals. Abnormal plumages occur more frequently in certain species than others, and can be common in small, genetically similar populations.

What is a 'phase'?

A phase (also sometimes called a 'morph') refers to the presence in a wild population of two or more distinctive plumage forms, which freely interbreed with each other. Examples include the pale phase and dark phase of Arctic Skuas, the red and grey forms of screech owls, and the 'bridled' form of the Guillemot. These are thought to be caused by a single genetic difference, rather like eye colour in humans, and do not mean that the individuals involved belong to different subspecies, even though they may look very different. Species with two different morphs

Albino Peacock

are known as 'dimorphic', while those such as male Ruffs, which have many different colour forms, are known as 'polymorphic'. Morphs are most widely found in two groups: owls and nightjars, in which up to one-third of all species may show distinct plumage phases.

What is 'sexual dimorphism'?

Simply the difference between a male and female of the same species, which may be expressed in size, colour, plumage features or a combination of all three.

Screech Owl – grey form

Screech Owl – red form

23

MOULT

Why do birds moult?

Because if they didn't, their plumage would get so worn and tatty they would have trouble finding food, coping with the elements, breeding – and ultimately surviving. Flight, in particular, is hampered by broken feathers and worn plumage. Birds also moult to adopt more showy plumage for courtship and breeding. This may simply be a brighter version of their usual plumage, or in some cases – for example, in many waders, grebes, divers and auks – a completely different-looking plumage just for the breeding season. Extreme examples include the African widowbirds and whydahs, whose males exhibit exotic breeding plumage with incredibly long tails, yet outside the breeding season become a nondescript streaky-brown, with short tails.

How do birds moult?

A new feather, growing from a follicle in the skin, gradually pushes out the old one. This usually follows a regular sequence within each feather group. So a passerine such as a thrush or warbler will usually moult its flight feathers from the innermost (and shortest) primary feather outwards, towards the longest feathers on the outside edge of the wing. Birds of prey have a more complex moult pattern, beginning with one of the middle primary feathers on the wing, with moult working in both directions. Many waterbirds, such as ducks, geese and swans, shed all their flight feathers at the same time, and are often unable to fly for several weeks until the new feathers have grown.

Long-tailed Widowbird outside breeding season (left) and in full breeding plumage (right)

How often do birds moult?

All adult birds moult at least once a year, and many species – including some waders (shorebirds), gulls and terns – do it twice, adopting a different appearance during the breeding season. Birds moult more frequently in their first year, sometimes shedding their feathers up to three times (from down to juvenile plumage, juvenile to 'first-winter' plumage, and finally into full adult breeding plumage the spring after they are born).

How long does moult take?

For most birds, a month or two, though for some passerines such as wagtails it can take up to 75 days. Long-distance migrants such as warblers and flycatchers tend to moult more quickly than sedentary species, as they need to grow new flight feathers in plenty of time to make the journey south after breeding.

Do all birds moult at the same time of year?

Not exactly, but the majority moult soon after the end of the breeding season. Moulting makes a bird more vulnerable to attack by predators, and may also make it less able (in some cases, completely unable) to fly. So most birds choose a time of year when there is plenty of food, dense foliage in which to hide, and no need to use up valuable energy in courtship or migration. Most migratory birds moult before they head off, taking advantage

of new flight feathers for the journey. However, a few (such as swallows and some birds of prey) wait until after they reach their destination.

What is 'eclipse' plumage?

It is simply a temporary plumage adopted by several families of birds, notably ducks, in which the male moults all his colourful feathers after courtship and mating, and resembles the duller female for several months.

Male Tufted Duck in 'eclipse' plumage

What are 'worn' and 'fresh' plumages?

These are terms that describe different stages in the moult cycle. 'Worn' usually applies to an adult bird towards the end of the breeding season, when the toll of finding food for hungry young has left its plumage looking very tatty. In contrast, once it has moulted, an adult acquires 'fresh' plumage, as does a juvenile following the shedding of its original downy feathers.

What are 'breeding' and 'non-breeding' plumages?

These describe the very different plumages adopted by some species during and outside the breeding season. However, because birds moult at different times of year, a bird may have acquired 'breeding' plumage well before courtship actually starts, or may attain its 'non-breeding' plumage while still feeding young. To make things even more confusing, many otherwise reputable field guides persist in using the even less accurate terms 'summer' and 'winter' plumages – despite the fact that in Europe, a Black-headed Gull may moult into its 'summer' plumage as early as December. To resolve the confusion, US ornithologists have proposed a new terminology, using the term 'basic' to replace 'non-breeding' or 'winter' plumage, and 'alternate' to replace 'breeding' or 'summer' plumage. Unfortunately this has not yet caught on in Britain, where the old, misleading terminology is still preferred.

What is the difference between 'immature' and 'juvenile' plumages?

'Juvenile' plumage is specifically defined as the first 'proper' plumage acquired at fledging: i.e. after the bird sheds its original downy feathers. 'Immature' is a less well-defined term, and simply means any plumage between juvenile and full adult – which, in the case of larger birds such as eagles and gulls, may cover half a dozen distinct plumages spanning a period of several years.

King of the wingers

The title of the world's largest flying bird is shared between the Andean Condor, weighing up to 15kg (33lbs), the Great Bustard, whose males weigh an average of 17kg (37lbs), and the African Kori Bustard, males of which may occasionally reach a weight of 18–19kg (around 40lbs).

The Andean Condor, one of the world's heaviest flying birds

SENSES

Which is the most important sense for birds?

In all but a very few cases, their sight. Birds tend to have proportionately larger eyes and more acute vision than other vertebrates, with most species depending on this to find food and/ or avoid predators. Hearing is also important, especially for songbirds and nocturnal species such as owls and nightjars. Some birds, especially seabirds and scavengers, also have a powerful sense of smell.

The 'ears' on the head of the Long-eared Owl are actually tufts of feathers

Which is the least important sense for birds?

Their taste. Most birds only have between 30 and 70 taste buds, whereas humans have about 9,000. However there are some exceptions: domestic chickens have around 250–350 taste buds, while parrots have up to 400 on their large, fleshy tongues. Birds do not generally distinguish between bitter and sweet tastes, though they can detect salt, which is an important supplement to many species' diet. Their lack of a refined ability to taste is probably because most birds swallow their food very rapidly, without chewing it as we do.

How well do birds see?

Birds' eyes, like those of other vertebrates, are complex organs: able to process visual information, and send signals to the brain that allow their owner to interpret the world around it. Without good eyesight, birds would find it almost impossible to fly at speed without bumping into things, let alone find food or dodge predators. It is generally assumed that birds see better than we do, and it is true that groups such as diurnal birds of prey have much better eyesight than humans. 'Better' means that they have a greater ability to see detail in distant objects; probably somewhere between two and five times the power of human eyes.

Peregrines have very acute sight, which they use when hunting

27

Do birds have binocular vision?

Most do not. Virtually all birds have eyes on either side of their head. This gives them a greater field of view than ours, but means they lack the combined focus of two eyes ('binocular vision') that enables us to judge perspective and distance. Having a greater field of view helps birds to find food and to avoid predators – indeed some, such as woodcocks and sandgrouse, have eyes that protrude from their sockets, so they can see more or less all around them. The only birds that have true binocular vision are hunters such as some hawks, eagles and owls. Having both eyes facing forward gives them a major advantage when hunting down their victim, though of course it does considerably reduce their field of view.

Can any birds turn their heads through 360 degrees?

Not quite. But owls can turn their heads through an arc of 270 degrees, thanks to specially adapted bones in their neck. This is necessary because, unlike other birds, owls

Owls can turn their heads through 270 degrees

are unable to turn their eyes in their sockets. However, many birds are able to see up to 320 degrees by a combination of turning their head and the position of their eyes. The Wryneck is another species that, as its name suggests, can twist its neck and turn its head through almost 360 degrees.

Bald Eagles have forward-facing eyes, giving them binocular vision

Can birds see colour?

Yes. In fact, most birds probably see colour rather better than we do, with richer tones and the ability to perceive some wavelengths of light which we cannot, such as the ultra-violet end of the spectrum. This helps them spot items of food, and also explains the prominence of colour in plumage and courtship displays. However, many nocturnal birds have poor colour vision, as this improves their sensitivity to low light levels. Penguins, too, have worse colour vision than other birds – probably because their environment is mainly monochrome!

Can birds see in the dark?

Yes, up to a point. No birds can see in total darkness, but many have much better night vision than we do, with the ability to see in very low light levels.

Barn Owls hunt using a combination of night vision and acute hearing

Several groups of birds, such as owls and nightjars, are predominantly nocturnal. Others, including tidal feeders such as wildfowl and waders, are often active after dark. The majority of migratory birds also prefer to travel by night.

How do owls find their prey?

By using a combination of good night vision, the ability to perceive tiny sounds made by moving creatures, and the kind of predictive ability that allows a car driver to travel a familiar route after dark. Some, such as the Barn Owl, hunt over open spaces where there are few obstacles. Others, like the Tawny Owl, have small home territories where they remain all year, allowing them to use their detailed knowledge of the terrain to hunt, even when they cannot see perfectly.

Which birds have the best eyesight?

It is often said that birds of prey such as the Peregrine have eyesight eight or ten times as acute as human vision. This is probably an exaggeration, though there is no doubt that their vision is considerably better than ours, and probably the best in some respects of any living creature.

The Peregrine's eyesight is at least eight times better than ours

Do birds have ears?

Yes, they have complex internal ears, just as we do, though they lack the fleshy external ear structures displayed by us and other mammals, and just have openings on the sides of their heads, under their feathers. Incidentally, the 'ear-tufts' on owls are not for hearing at all, and may have evolved for camouflage.

How well do birds hear?

By and large, pretty well. Although vision is the best developed sense amongst birds, most also have fairly sharp ears. For songbirds, hearing is obviously

critical, enabling them to find a mate. Songbirds can also hear more complex sounds than we can, by 'slowing down' the sequence of notes in order to interpret the signal. Owls also have incredible aural ability: having one ear positioned slightly lower than the other gives them 'binaural hearing', enabling them to pinpoint the exact position of their prey, even in total darkness.

Chaffinches can hear sounds far higher in pitch than we can

What range of sounds can birds hear?

Birds as a whole can hear sounds ranging from about 35 hz to just below 30,000 hz (the measurement of sound frequency, in which the greater the number the higher the pitch). Our range, at between 20 and 20,000 hz, is slightly lower. However, individual species have a far more restricted hearing range, with most ranging between 100–200 and 3,000–18,000 hz, though some songbirds such as the Chaffinch can hear sounds as high as 29,000 hz.

Whopping wings

The greatest wingspan probably belongs to the Wandering Albatross, with a recorded length of 3.6 metres (almost 12 feet) – though some individuals of this species and the Royal Albatross may have even longer wings. For total wing area, the title goes to the Andean Condor, whose wingspan reaches 3 metres (almost 10 feet). But today's giants are dwarfed by a prehistoric condor in the genus *Teratornis* ('monster bird'), which was also the largest flying bird that ever lived. This titan of the Pleistocene era (more than one million years ago) had a wingspan of over 7 metres (23 feet) and weighed about 80kg (175lbs). Its remains have been found in the La Brea tar pits in Los Angeles, California.

The Wandering Albatross has the largest wingspan of any bird

How well can birds smell?

Some can smell pretty well. Seabirds such as storm-petrels can sniff out food sources as much as 25km (16 miles) away. Taking advantage of this, ocean-going birders spread a foul-smelling concoction of fish guts known as 'chum' on the surface of the water, in order to attract the birds. Scavenging species such as Turkey Vultures also use smell to locate rotting carrion, which may be hidden beneath the forest canopy – though the unrelated vultures of the Old World do not have a well-developed sense of smell, and search for food on the open savannas primarily by sight. Other champion sniffers include nocturnal species such as the kiwis of New Zealand, which locate their underground prey using nostrils at the tip of their bill.

The Painted Snipe uses its long bill to probe into mud for food

What about birds' sense of touch?

This has been studied far less than the other senses, so there is still much to be discovered. However, we do know that some birds locate food using sensitive nerve endings in their bill-tip and/or tongue: examples include waders, such as snipe, that probe into mud; and woodpeckers, which winkle out food from beneath tree bark.

How do birds keep warm?

In various ways. Most fluff out their feathers to trap an insulating layer of air. Some also huddle together in winter roosts, taking advantage of their collective body temperature. Food is a vital factor: a hungry, underweight bird is far more likely to perish from cold than a fat healthy one.

Why don't waterbirds get frostbite?

Waterbirds such as ducks and geese face a special challenge: keeping warm while standing on ice. They do so by reducing the blood supply to their legs and feet, thus minimising heat loss. They also have fewer nerve endings in their feet, so they do not feel the effects of the cold.

Turkey Vultures have a highly developed sense of smell

Small is beautiful

The world's smallest bird is the Bee Hummingbird of Cuba, which weighs between 1.6 and 2 grams (1/15th to 1/12th of an ounce) and measures just 5.7cm (2.25in) long, half of which is bill and tail.

The title of world's smallest passerine belongs to both the Black-capped and the Short-tailed Pygmy-Tyrants of Costa Rica, which are only 6.5cm (2.5in) long. The world's smallest seabird is the Least Storm-petrel, which breeds around the Gulf of California. It is just 13–15cm (5–6in) long, and weighs less than an ounce.

The world's smallest raptors are the falconets of South-east Asia, including the Collared, Black-thighed, White-fronted and Philippine, all of which range from 14 to 18cm (5.5–7in) long, and weigh as little as 30g (1.05oz).

Finally, the world's smallest flightless bird is the Inaccessible Island Rail,

The Bee Hummingbird – the world's smallest bird

found on the South Atlantic island of the same name. It is just 13cm (5in) long and weighs a mere 34g (just over one ounce).

Collared Falconets are one of the world's smallest raptor species

SLEEP

How do birds sleep?

Much like us, they drop off in all sorts of different ways: from brief 'cat-naps' to extended slumbers. But it is a myth that birds doze off with their head tucked beneath their wing. In fact most sleep either with their head turned back and tucked beneath their shoulder feathers, or with their head slumped back onto their shoulders in a hunched position.

Where do birds sleep?

Birds need to find safe places to sleep. So perching birds usually sleep on a hidden twig or branch (or in the case of woodpeckers, clinging to a tree-trunk), waterbirds either sleep on the water or stand on nearby ground (especially islands), and game birds hide in dense vegetation to avoid predators. Swifts, the most aerial of birds, sleep on the wing, rising high into the atmosphere to do so. Many birds gather in roosts, where they can be safer from predators and steal some warmth from others, especially in winter. Thus a relatively solitary species such as the Pied Wagtail may gather in flocks of several hundred on cold nights.

Do birds sleep standing up?

Yes: either on the ground, or perched on a twig or branch. Many – especially songbirds, flamingos and waders – will stand alternately on one leg or the other, presumably to conserve body heat in cold weather.

Flamingos sleep by tucking their head beneath their feathers

When do birds sleep?

Like us, most diurnal birds sleep during the hours of darkness, when it is difficult or impossible to find food. Most nocturnal species such as owls and nightjars do the opposite, and roost during daylight hours. However, for many groups, especially waterbirds such as ducks and waders, sleep periods coincide with high tides when food is unavailable.

Why don't sleeping birds fall off their perches?

It is often supposed that perching birds contract their tendons automatically, to prevent them from falling off their perch, even when fast asleep. However, recent findings suggest that they may simply be very good at keeping their balance.

Do birds hibernate?

For centuries it was believed that many birds hibernated, with summer visitors like Swallows thought to spend winter in the mud at the bottom of ponds. But this was conclusively disproved by careful observation and ringing studies in the 19th and 20th centuries. Then, in the 1940s, an incredible discovery was made: a Common Poorwill, a type of American nightjar, was found in a torpid state in a rock crevice in California. Studies showed the bird was effectively in a state of hibernation, achieved by reducing its body temperature from the normal 40°C to just 10°C. It is now known that poorwills can maintain this

The Common Poorwill is one of the few birds that can lower its body temperature to enter a form of hibernation

state for up to 100 days. Other birds such as some hummingbirds and swifts can also shut down most of their body mechanisms to achieve a state of torpor, but for much shorter periods.

MORTALITY

How long do birds live?

Anything from a few hours to almost as long as us, depending on the kind of bird and the hazards it faces. Many birds die very early in their lives, due to starvation or attack by predators. If a bird survives its first year, then lifespan varies between species: most songbirds live for between two and 10 years, waders from about five to 10 years, and raptors from five to 20 or even 30 years. Parrots and seabirds are among the longest lived, with many species regularly topping 20 years and a few individuals breaking the half-century barrier. The longest surviving birds are those in captivity, which have no predators to face and an unending supply of food, and may possibly live for as long as 100 years. But very few wild birds will ever reach 'old age'.

Many garden birds, like this Robin, fall victim to domestic cats

What proportion of a bird's population dies each year?

Usually between one-third and two-thirds, though this varies from species to species and depends on environmental conditions such as winter weather, or food shortages. Seabirds tend to have very low annual mortality rates (five to ten per cent), while amongst small passerines up to three-quarters of a population may die each year.

How do birds die?

In many different ways: some natural, others due directly or indirectly to human agency. Disease, lack of food and predation are the three biggest natural killers, while non-natural deaths arise from factors such as shooting, deliberate or accidental poisoning, and collisions with buildings or motor vehicles. For garden birds, one of the biggest causes of death is predation by

domestic cats. It is estimated that each year, cats are responsible for anything from 28 to 75 million bird deaths in the UK and as many as 118 million in the US.

Do birds suffer from the same diseases as humans?

Yes, many birds suffer from avian forms of common human diseases. These include influenza, tuberculosis, botulism and salmonella poisoning. Occasionally, diseases can spread from birds to humans, as in the bacterial disease psittacosis, and recent outbreaks of West Nile Virus and 'bird flu', all of which can be potentially fatal to human beings.

Why do you rarely see a dead bird?

Considering that millions of birds die each year, it is amazing how few dead ones we come across. Many die out of sight of human beings, while others are killed and eaten, leaving little or no evidence. Decay and decomposition can also be very rapid, especially in warm climates.

INTELLIGENCE

How intelligent are birds?

'Bird-brained' is a very unjust insult, since birds are mostly pretty smart. However, many signs of apparent intelligence in birds, such as the ability to find food or navigate over long distances, cannot really be compared with human brainpower, since much of it is simply 'programmed' behaviour. Nevertheless, birds can be trained to solve quite complex problems, and some have the ability to use tools (see p.38).

How do birds learn to do things?

Many basic life skills such as finding food are learned by watching and imitating their parents, though innate instinct also plays a part. Thus male birds kept in isolation from others of their species are still able to sing, though not as well as if they had been able to hear other singing males. Birds also use a process of trial and error: if they eat a particular insect or plant that makes them ill, they are unlikely to do so again. Finally, birds are also capable of learning through observation of cause and effect, such as when Herring Gulls learn to drop a mollusc onto a hard surface in order to get at its contents, or Blue Tits discover how to peck through the foil tops of milk bottles to reach the cream below.

Parrots are amongst the most intelligent of all birds

Can birds count?

Not as well as we might think. The ability to count has often been ascribed to pets such as dogs, horses and intelligent birds such as parrots. However, experiments have shown that this is normally either a deliberate hoax by the animal's owner, or reflects an unconscious reaction on the owner's part – such as relaxing when the animal reaches the 'correct' number.

Cygnets learn to feed from their parents

Can birds solve complex problems?

Crows, like this Japanese Crow, use their intelligence to find food

It used to be thought that birds were less able to solve problems than, for example, rats and squirrels. However, this was mainly based on laboratory tests using pigeons, which are perhaps not the smartest of birds. Further studies have shown that some birds, especially crows and parrots, can often handle complex and multi-faceted tasks, such as navigating a maze or doing steps in a particular sequence to gain a reward. But perhaps the most extraordinary example of wild birds' intelligence has been observed in Japan. In certain Japanese cities, crows deliberately place walnuts on the surface of a road when the traffic lights are on red. They wait until the lights change, watch as a car or lorry crushes the nut, and then – when the light changes back to red – swoop down to pick up the kernels and eat them.

Can birds use tools?

Quite a few do, and usually to find food. The most famous example is the Woodpecker Finch of the Galapagos, which habitually uses a small twig or cactus spine to provoke an insect to emerge from a crevice or to physically prise it out. The New Caledonian Crow goes one stage further by fashioning its own tool from a twig in order to prise grubs out of a tree trunk. Some birds that fish for a living have also learned to improve their success rate by using bait. For example, the Green Heron

Gulls often smash shellfish against hard rocks to get at the contents

How did Blue Tits learn to open milk bottle tops?

Between the First and Second World Wars, milk delivery companies in Britain introduced aluminium foil tops to milk bottles, helping to keep the milk fresh when left on the doorstep. It did not take long for one enterprising species, the Blue Tit, to discover how to get at the tasty cream inside the bottle, by pecking a hole in the foil. This phenomenon was first recorded in the 1920s, and by the 1950s had spread throughout the British Blue Tit population. The speed at which the habit spread led some scientists to claim that this was an example of 'Lamarckian inheritance', in which a trick learned by the parent can be

has learned to drop insects or small pieces of bread on the surface of the water to attract fish, which it can then seize with its dagger-like bill. Several species use hard objects to obtain food: either dropping stones onto eggs to break them (as one African population of Egyptian Vultures does to Ostrich eggs); or by dropping an item of food onto a hard surface (gulls dropping shellfish onto concrete, or Lammergeiers releasing bones from a great height). This behaviour has passed into legend: the Greek playwright Aeschylus was reputedly killed by an eagle dropping a tortoise on his bald head; the 'eagle' is more likely to have been a Lammergeier (though legend doesn't recall what grudge it held against playwrights).

Blue Tits learned to raid milk bottles to get the energy-rich cream

passed down to its offspring via its genes. The truth was perhaps even more astonishing: because Blue Tits live in family groups, the offspring were watching their parents and finding out how to pierce the foil tops by a combination of observational learning and trial and error. Interestingly, although some individual Robins also learned to get into the milk bottles, the habit did not spread through their population, and soon died out. This was because Robins are mainly solitary creatures, so there was little opportunity for the skill to pass from one generation to the next. After complaints from customers, the dairy companies first tried strengthening

The aptly named lovebirds are highly affectionate

the foil tops, but the birds simply pecked harder! Eventually, a change in packaging design and shopping habits led to a decline in the doorstep milk delivery, and consequently the end of a free meal for Blue Tits.

Do birds have feelings?

Watching Rooks tumbling through the sky, a songbird singing its heart out, or a mother standing forlornly over her dead chick, it is easy to assume that birds have feelings very similar to our own. But to use words like 'happiness' or 'sadness' in relation to other members of the animal kingdom raises many questions. Lots of creatures indulge in what can be described as 'play', 'courtship' or 'aggression', but we cannot suppose that when doing so they feel the same as we do – or even have the capacity to feel emotion at all. Scientists divide into two camps on this question: one side sees behaviour in a purely mechanistic sense (e.g. young creatures playing is a way of learning to hunt); the other allows the notion that birds and other animals do have emotions and feelings which, while not exactly the same as ours, have their roots in the same biological imperative. The latter approach has most eloquently been described in *The Minds of Birds*, by the late American ornithologist Alexander Skutch, in which he concludes that "birds' mental capacities have been grossly underestimated… their minds are amongst nature's greatest wonders." Many people would agree with him.

Golden Oldies

The world's longest-lived wild bird is a female Layson Albatross ringed as an adult in 1956, when she was at least five years old. At the time of writing, she is at least 64 years old. In the UK, a Manx Shearwater ringed as an adult in 1953 on Copeland Island, County Down, was recovered in 2003, making it at least 55 years old, while a Fulmar ringed on Orkney in the 1950s was recovered 44 years later, aged about 50. The longest-lived bird in captivity was a Sulphur-crested Cockatoo named Cocky, which died in London Zoo in 1982, aged at least 80.

Sulphur-crested Cockatoos are one of the longest-lived of all birds

2 · WHERE DO BIRDS COME FROM?
Evolution and classification

HOW DID BIRDS EVOLVE?

Like every other creature on this planet, birds got here by evolution – a process discovered by 19th century British naturalists Charles Darwin and Alfred Russel Wallace – in which individuals with genetically determined traits most suited to survival in their particular environment pass on these traits to their offspring. These did not appear as the result of any preordained plan (hard as some may find this to accept), but are simply the product of tiny genetic changes accumulated over a vast period of time. Evolution is an ongoing process, with new species continually evolving while others become extinct. So all life is in a state of constant flux.

A Velociraptor

Did birds evolve from dinosaurs or reptiles?

For many years the jury was out as to whether birds had descended from fully-fledged dinosaurs or some other branch of prehistoric creatures such as reptiles. However, in recent years the discovery of many new fossils has finally settled the question, and it is now widely accepted that birds descended from a group of dinosaurs known as theropods - mainly smaller, carnivorous dinosaurs that include the intelligent, rapid and somewhat bird-like velociraptors made famous by Steven Spielberg's classic film Jurassic Park.

Archaeopteryx fossil

What is taxonomy?

Taxonomy is the study of the evolutionary relationships between different populations of animals or plants, and their classification into a great zoological filing cabinet of orders, families, genera, species and so on. It is also known as systematics.

So what is taxidermy then?

Taxidermy is the craft (some would say art) of preparing the skins of birds, mammals and occasionally fishes for display in a museum or private collection. Put more crudely, it is all about stuffing animals and has nothing to do with classification.

A stuffed Water Rail in a display case

What is the earliest known fossil bird?

Most scientists seem to agree that one of the first true birds was the celebrated *Archaeopteryx*, specimens of which were discovered in Bavaria during the 19th century. Dating from roughly 150 million years ago, *Archaeopteryx* has some traits of both reptiles and birds, having a reptilian skeleton and teeth, but also well-developed, flight–capable feathered wings. However, other fossil discoveries suggest that *Archaeopteryx* may have been an evolutionary 'dead end', and there are several other contenders for the title of the world's first bird.

43

How do we know how to classify birds?

All forms of taxonomy are based on a combination of scientific knowledge and inspired guesswork. So two different taxonomists may disagree – especially when it comes down to the tiny differences that separate one species from another, or deciding where to draw the line between different genera. Several centuries ago, birds were classified in terms of what they ate, or their habits. Thus Osprey and Kingfisher were grouped together as 'fish-eating birds', while eagles and owls were 'raptors', and thought to be related to one another. Later on, by studying the insides of birds as well as their outsides, scientists created a more accurate system of classification, which more closely reflects the way birds actually evolved. Today, genetic analysis reveals the finer detail of relationships between species.

Kingfishers (above) and Ospreys (below) were once thought to be related because of their diet of fish

So is today's classification of birds the 'true' one?

Almost certainly not. The Holy Grail of taxonomists is to trace the true evolutionary path of each species, and thus the order in which all today's species descended from a common ancestor. This process is known as phylogeny. However, taxonomists can never be absolutely sure that any evolutionary path is the true one, although DNA and voice studies bring us ever closer to this goal.

Owls, like this Snowy Owl, were once classified with hawks and eagles as raptors

What is an 'order'?

In the family tree of classification, an order is the main taxonomic category between class (e.g. birds, mammals, insects) and family (e.g. parrots, owls, buntings – see below). There are somewhere between as few as 23 and as many as 39 different orders of birds, depending on whose opinion you follow. The scientific name of an order always ends in '-iformes'. Orders range in size from those containing a single species, such as Opisthocomiformes (the unique Hoatzin of South America), to the largest order of all, Passeriformes, which contains almost 6,000 species – more than half the world's total bird species.

Taking orders

The largest order of birds is without question the Passeriformes (passerines), which includes almost 6,000 of the world's extant bird species – well over half the total.

The next largest are the Apodiformes (swifts and hummingbirds), with about 440 species, followed by the Charadriiformes (waders, gulls etc.) and Psittaciformes (parrots, macaws and cockatoos), with about 350 species each.

There are several candidates for the smallest order of birds, depending on whose taxonomy you follow. Until recently, most authorities recognised only one order with a single species: Struthioniformes (the Ostrich). However, the Hoatzin, a peculiar South American bird long considered a member of the order Galliformes, has now been given its own unique order, the Opisthocomiformes. To further confuse things, some authorities now recognise two separate species of ostrich.

Corn Bunting – Passeriformes

Black-headed Gull – Charadriiformes

Swift – Apodiformes

Blue-and-yellow Macaw – Psittaciformes

What is a 'passerine'?

A passerine is a member of the order Passeriformes. Because this order is so large, it has been sub-divided into two 'sub-orders', the oscine and sub-oscine passerines, split according to the structure of their vocal parts. The vast majority of passerines, including sparrows, finches, warblers, tits and larks, are oscine passerines; the sub-oscines include the antbirds and cotingas, and are mainly confined to South America.

Spangled Cotinga – a sub-oscine passerine

Are passerines the same as songbirds?

A 'songbird' is a member of the oscine passerines, by far the largest grouping in the order. All passerines that regularly occur in Britain and Europe are oscine passerines, so can be accurately described as songbirds. However, in North America, the tyrant-flycatchers (including kingbirds, phoebes and flycatchers) are sub-oscine passerines and so are not true songbirds, though all other North American passerines, from warblers and thrushes to sparrows and larks, are.

What is a 'raptor'?

Raptor is generally used as a synonym for 'bird of prey', and usually describes members of the orders Accipitriformes and Falconiformes (eagles, hawks, buzzards, falcons etc.), and sometimes Strigiformes (owls). It is not always easy to define why a bird is or is not a raptor: not all raptors eat meat (e.g. the Palm-

Rough-legged Buzzard – a raptor

nut Vulture); several are scavengers (e.g. Old World vultures); and there are some birds that behave like raptors but are not (e.g. shrikes).

We also exclude birds such as gannets, even though they feed exclusively by killing other animals (in their case, fish).

What is a 'wader'?

The term 'wader' is used for species that belong to any of various families in the order Charadriiformes, such as plovers, sandpipers, avocets and stilts, and phalaropes. It does not apply to all families in this order though. Gulls, terns, auks and skuas are not waders. In North America, waders are generally referred to as 'shorebirds', the term 'wader' being reserved for other long-legged wading birds such as herons and egrets.

Pacific Golden Plover – a wader

What are 'wildfowl'?

Wildfowl (or waterfowl as they are known in North America) are members of the order Anseriformes; ducks, geese and swans.

Bar-headed Goose – a member of the Anseriformes, often known as wildfowl

The Gannet is the largest seabird in the northern hemisphere

What is a 'seabird'?

Broadly speaking, the term 'seabird' refers to any bird that spends much of its life at sea. However, it is usually confined to members of the following families: tubenoses (albatrosses, shearwaters and petrels), gannets and cormorants, gulls and terns, skuas, auks, penguins, frigatebirds and tropicbirds. Other groups, including divers, grebes, phalaropes and certain species of ducks, are not usually referred to as 'seabirds', even though they may spend much of their time at sea.

What is a 'family'?

In the table of classification, family is the level between order and genus. Families may contain a single species (e.g. Hoatzin), a dozen or more (e.g. albatrosses) or several hundred (e.g.

hummingbirds, parrots, pigeons and doves). The scientific name of a family always ends in '-idae' (e.g. Fringillidae – finches).

What is a 'genus'?

Genus (plural: genera) falls between family and species. Effectively it is a convenient grouping of closely related species, although the placing of a species within a particular genus is not always clear-cut, with the boundaries often shifting. So one species may be moved between genera, or one genus may be split into two or more genera – or combined with another to form a single one. For example, some scientists now believe that the two Old World warbler genera *Hippolais* (e.g. Melodious Warbler) and *Acrocephalus* (e.g. Reed Warbler) should be combined into one.

Family affairs

The largest family of birds is the tyrant-flycatchers of the New World, with between 340 and more than 500 species – the exact number depending on whose authority you recognise. Of the non-passerines, the largest family is either the parrots or the hummingbirds, both of which have about 320–350 species, closely followed by pigeons and doves, with just over 300 species.

There are many candidates for the smallest family of birds. According to James Clements' latest checklist of the world's birds, 16 non-passerine families and seven passerine families contain only a single species (known as monotypic families). They are: Ostrich, Emu, Hamerkop, Shoebill, Osprey, Secretary Bird, Hoatzin, Limpkin, Kagu, Sun-bittern, Crab Plover, Ibisbill, Magellanic Plover, Plains-wanderer, Oilbird and Cuckoo-Roller (non-passerines); and Sharpbill, Hypocolius, Palmchat, Wallcreeper, Bristlehead, Olive Warbler and Bananaquit (passerines). However, this is constantly changing; the latest taxonomy suggests that the Bearded Tit, once lumped with the parrotbills, may actually be in a monotypic family.

The Hoatzin is in its own unique family, as it is not closely related to any other species of bird in the world

Crossbills (male left, female right) have a very complicated taxonomy

What is a 'species'?

This used to be a nice simple question to answer, but thanks to recent advances in taxonomy, it's now, frankly, a bit of a nightmare. Until recently, a species was defined as a group of birds which freely interbreed with each other, but generally do not breed with other groups of birds. In practice, this meant that the birds we normally recognise as being different from each other are different species. However, some apparently distinct species, such as the larger gulls, may appear to be separate species in one part of their range, while interbreeding in another. Moreover, populations which until now have been considered as making up a single species, such as the Common (Red) Crossbill, may in fact be a number of separate species – even though they are virtually indistinguishable in the field (at least to us humans). It

is important to understand that no species is fixed forever in evolutionary terms; populations constantly have the potential to eventually evolve into a separate species.

What is a 'subspecies'?

A subspecies – or 'race' – is a convenient classification of particular populations of a species into different categories. It is usually applied when a population of birds in one area differs from a population of the same species in another area, but these differences are considered too minor to warrant classifying them as separate species. A good example is the division of *Motacilla alba* into two subspecies: 'Pied' Wagtail (subspecies *yarrellii*), breeding in Britain; and 'White' Wagtail (subspecies *alba*), breeding on the European mainland. They can generally be told apart by subtle but still

Despite differences in appearance, Pied Wagtail (left) and White Wagtail (right) are two races of the same species

noticeable plumage differences: for instance the Pied Wagtail has a much darker back than the White.

What is the difference between a 'race' and a 'subspecies'?

None: they are simply synonyms.

Do subspecies ever 'graduate' to becoming full species, and vice versa?

Yes – by 'lumping' and 'splitting' (see page 52).

What are 'binomials'?

The system of binomial nomenclature was created more than two centuries ago by the Swedish scientist Linnaeus, yet is still in use today. Linnaeus made the simple but revolutionary breakthrough of giving each species a unique scientific name, in two parts. So the House Sparrow has the scientific name *Passer domesticus*, in which the first part, *Passer*, refers to the sparrow genus, and is shared with two dozen or so other species; while the second part, *domesticus*, indicates that this is the House Sparrow (as opposed to the Tree Sparrow, *Passer montanus*). Although neither the generic nor the specific name is necessarily unique, the combination of the two always is.

The House Sparrow's scientific name is *Passer domesticus*

What's the point of binomial nomenclature?

By giving each species a unique scientific name, we can always be certain about which species is being referred to. This is especially useful when different species share a common English name, such as the two unrelated species known as 'Black Vulture', one of which is found in the Old World (*Aegypius monachus*), and the other in the New (*Coragyps atratus*). It is also very helpful when referring to a species that has different English names: for example the bird known in Britain as Arctic Skua, and in North America as Parasitic Jaeger, but by scientists everywhere as *Stercorarius parasiticus*.

What about 'trinomials'?

Trinomials are simply an extension of the binomial concept in order to differentiate between different subspecies (or races) of a particular species. Thus the various races of the Yellow Wagtail *Motacilla flava* are distinguished from each other by adding a third word to each scientific name. So the race breeding in Britain is known as *Motacilla flava flavissima*, while that breeding in central Europe is known as *Motacilla flava flava*. The repetition of the specific name as that of the subspecies indicates that the latter is the so-called 'nominate race' – i.e. it was the first to be discovered and named by scientists.

What do 'lumping' and 'splitting' mean?

These terms are birders' slang for what happens when two formerly distinct species are reclassified as a single one (lumping), or two subspecies are given full status as separate species (splitting). In recent years splitting has become far more frequent. For example, until the mid 20th century Pink-footed and Bean Geese were considered to be two races of the same species. Then it was found that they were reproductively isolated from one another, and so they were 'split' into two full species. Some authorities have since done the same with the various types of Brent Goose (Dark-bellied, Pale-bellied, Black Brant etc.).

Why has lumping declined?

In recent years, 'lumping' – the joining together of two or more species into one – has become less frequent. This is partly because of the adoption by some authorities of the Phylogenetic Species Concept, as opposed to the traditional Biological Species Concept (see opposite); and also because as we get to know more about bird distribution, we have discovered that what we once thought were races are in fact reproductively isolated from one another, and are therefore separate species.

Pink-footed (top left) and Bean Goose (top right) used to be considered races of the same species

What is the 'Biological Species Concept'?

The Biological Species Concept (BSC) uses the traditional way of defining species by the fact that they do not breed with other, similar species. But there are several problems with this. First, it is impossible to know whether two supposed species, whose ranges do not overlap, would breed with each

Three races of Brent Goose: dark-bellied (centre left), light-bellied (bottom left) and black brant (bottom right)

other if their paths did cross. On the other hand, we also cannot tell whether different subspecies, whose ranges also do *not* overlap, would breed with each other if they got the chance, or would keep to their own kind. So as a working definition, the BSC creates as many problems as it solves.

What about the 'Phylogenetic Species Concept'?

To try to solve these problems, some taxonomists have come up with a new system of classifying birds, the Phylogenetic Species Concept (PSC). They propose that we try to identify the smallest population of a bird that has a unique set of characteristics, and consider this population to be a separate species. So, for example, every distinct race of the Blue Tit (at least nine exist) would be classed as a separate species, provided we could show that its members share a definable set of characteristics.

How do scientists tell species apart using the Phylogenetic Species Concept?

With great difficulty! Any new idea is bound to have teething troubles, and this one is no exception. Like more traditional taxonomists they use a number of ways: from examining the internal structure and external features of a bird, to analysing its song, and using DNA analysis to compare populations.

What would happen if we adopted the Phylogenetic Species Concept?

Taken to its logical conclusion, this would create taxonomic anarchy. Instead of 300 or so regularly occurring species in Britain there might be as many as 1,000 – many of them only distinguishable from each other either by analysing their sound recordings or close examination in the hand. Meanwhile the 'world list'

The Blue Tit has several distinctive subspecies, which may in the future be treated as separate species

Common Tern (top) and Arctic Tern (bottom) are 'sibling species'

of birds would increase from around 10,500 species currently to more than 20,000 species. Although twitchers might initially be thrilled by their ever-expanding lists, they would soon become deeply frustrated by their inability to make firm identifications. Novice birders would be even more baffled, and might give up altogether. And it is not just birders who are affected by this new approach: the familiar Pipistrelle Bat was recently split into two separate species, the Common and the Soprano Pipistrelles, based on the fact that one echo-locates at a frequency of 45 kilohertz, while the other does so at the higher frequency of 55 kilohertz.

What are 'sibling species'?

Sibling species are two or more closely related species that look similar, but do not normally interbreed, even when their ranges overlap. A good example is Common and Arctic Terns.

What are 'cryptic species'?

They are two species that so closely resemble each other that it is almost impossible to tell them apart in the field – even though they are reproductively isolated and constitute 'good' species. For example, two American sparrows, Nelson's Sharp-tailed and Saltmarsh Sharp-tailed, were until recently considered to be one species, but several subtle differences in plumage, song and habitat have now led them to be split into two.

Darwin's Finches helped Charles Darwin work out his theory of evolution

What is 'adaptive radiation'?

Adaptive radiation occurs when a single group of closely related organisms, descended from a common ancestor, evolves into many different forms, usually in a relatively short time. This generally occurs when they colonise a new area – such as an island or even a new continent – and evolve to fill the vacant niches available. The most famous example is Darwin's finches of the Galapagos, 15 or so species which, despite their very different appearances, all share a fairly recent common ancestor. The study of these birds helped Darwin to form his theory of evolution by natural selection.

Wacky races

The species with the largest number of subspecies is the Island Thrush, with 51 races currently recognised, many of them confined to tiny islands in the Pacific Ocean. The honour used to go to the Golden Whistler of South-east Asia, with 64 recognised races, but this has now been split into eight different species, each with a number of different subspecies.

What is 'convergent evolution'?

This occurs when two completely unrelated species or groups evolve similar characteristics, due to the evolutionary pressures of sharing a similar environment. Thus the Moorhen looks superficially like a duck or a grebe, because it shares their habitat and lifestyle, while its close relative the Corncrake does not. Another good example is the extraordinary resemblance of the African longclaws (related to Old World pipits and wagtails) to the completely unrelated American meadowlarks (related to New World orioles and blackbirds).

The Corncrake (top) and Moorhen (bottom) are both rails, but one has evolved to live on land and the other on water

Are birds still evolving?

Every living species is still evolving – the process never stops. And they are doing so much faster than we once thought. It used to be assumed that birds (and other creatures) evolved over thousands or tens of thousands of years – a far slower process than we could witness in our short lifetimes. But this assumption was shattered when studies of Darwin's finches on the Galapagos revealed evolutionary changes occurring over a few generations, covering only a decade or so. This was due to a major change in the birds' environment, brought about by the weather system known as 'El Niño'.

Darwin's finches are a classic example of rapid evolutionary change

Are we still discovering new species?

Yes: both by the process of 'splitting' (mainly using DNA analysis) and because every now and then a species entirely new to human knowledge is actually found in the field, having previously been undiscovered or overlooked. During the past decade or so new species are being identified at a rate of between three and 12 a year, mostly from South America and South-east Asia – recent discoveries include the Alta Floresta Antpitta from Brazil, the Cebu Hawk-owl from the Philippines, the Sierra Madre Ground Warbler – also from the Philippines – the Wakatobi Flowerpecker from Indonesia, and the Bahian Mouse-coloured Tapaculo from Brazil.

One species recently discovered by scientists is the Wakatobi Flowerpecker

3 • HOW MANY BIRDS ARE THERE?
Population

NUMBERS

How many species of bird are there?

That depends on who you ask, and when. Until the mid 20th century the number of bird species stood at about 8,600. Then there was a revolution in taxonomy – the scientific classification of living things (see Chapter 2). Using new techniques, scientists analysed the genetic make-up of bird specimens and discovered that many so-called races (or subspecies) of birds might in fact be 'proper' species. Further fieldwork confirmed this, and birds such as the Hooded Crow (previously thought to be a race of the Carrion Crow) were upgraded to full species status. Meanwhile, there was a steady trickle of entirely new birds being discovered, mainly in South America, equatorial Africa and South-east Asia. So by the turn of the millennium, when ornithologist James Clements published the fifth edition of his world checklist, he recognised more than 9,800 species – an increase of about one-seventh

Hooded Crow

on the old figure. The latest BirdLife International checklist, compiled with the help of Lynx Edicions, publishers of the monumental *Handbook of the Birds of the World*, recognises 10,426 species, an increase of about six per cent in just 15 years – but this is bound to increase further as more and more species are discovered or 'split' from their close relatives.

How does this compare to other animals?

There are roughly twice as many different species of bird as there are mammal (just over 5,000), and there are also more birds than amphibians (roughly 6,000) and reptiles (about 8,000). However, the 10,400 or so species of bird are easily outnumbered by fish (roughly 30,000 species), and positively swamped by insects, of which there may be as many as six million – with many still awaiting discovery.

How many species of bird have ever existed?

Estimates for the number of bird species to have lived since *Archaeopteryx* (thought by scientists to have been one of the first birds – see Chapter 2) range from roughly 150,000 to more than 1.5 million. This is despite the fact that only about 900 extinct birds have been identified from fossil remains,

and is based on palaeontologists' theories of the rate at which species become extinct over time. However, the latest thinking is that today's 10,400 or so species represent about six per cent of those that have ever lived, giving a likely total of about 170,000 species – roughly 16 times as many as exist today.

How many individual birds are there in the world?

In 1951, British ornithologist James Fisher estimated that there were about 100 billion individual birds in the world. If the same is true today, then birds outnumber human beings by about 15 to one. Amazingly, nobody appears to have updated this estimate since then. This is despite considerable advances in what we know about the world's birds, and the many population changes that have occurred.

Starlings gather in huge flocks before roosting – some over a million strong

Breeding frenzy

The world's most numerous wild bird is the Red-billed Quelea. This small seed-eating bird of the weaver family is found in the dry savanna region of sub-Saharan Africa. It occurs in flocks of at least 100 million, and its total population has been estimated at several billion individuals. Quelea colonies cover several hundred hectares, and contain up to ten million nests. Their success can be attributed to their breeding rate: pairs breed up to four times a year, and take less than four weeks from laying eggs to fledging their chicks. Queleas are a major agricultural pest, and millions are killed by aerial spraying each year.

CHANGES

Peregrine numbers have recovered since harmful chemicals were banned

Do bird populations change over time?

Like all creatures, birds are subject to a wide range of external influences, from habitat loss to persecution and climate change. These affect the breeding success of individuals and, over time, the population as a whole. Within a single human lifetime, often much less, we may witness major rises and falls in bird populations. So raptors such as the Peregrine declined rapidly in Europe and North America during the 1950s and 1960s, because of the use of agricultural chemicals such as DDT, but quickly recovered once these pesticides were banned. Some others have fared much better by adapting to living alongside people, feeding on our waste products and nesting on our buildings. Worldwide, populations of some species such as the Cattle Egret have positively exploded, thanks to their pioneering spirit and their handy ability to thrive alongside humans.

What causes bird populations to rise or fall?

The major causes of population declines are habitat loss, persecution and pollution. Habitat loss is the biggest factor in the tropics, where illegal logging and the clearing of forests for agriculture have brought hundreds of species to the brink of extinction. Persecution is generally in decline, especially in Britain and North America, though it is still a problem in the developing world and around the Mediterranean, where many millions of migrating birds are shot each year. Pollution can have a major effect in a small area, for example when an oil tanker runs aground (such as the 1989 Exxon Valdez disaster in Alaska, which killed at least 250,000 seabirds). In the future, the effects of global climate change may have profound consequences for many of the world's birds.

An oiled Guillemot – a victim of pollution

How do we measure these rises and falls in populations?

Britain and North America have a long tradition of amateur birdwatchers carrying out census and survey work. Over time they have helped give us a picture of the status, distribution and numbers of breeding and wintering birds, so we can detect changes in range or population very soon after they occur. Without the data produced by surveys such as the BTO's *Bird Atlas* we could never be confident that these changes are actually happening, nor could we measure them so accurately.

Seabird colonies are easier to survey than other breeding species, as the birds are easily visible

How accurate are our estimates of populations?

Inevitably they are a combination of hard facts and educated guesswork. Some species, such as colonially nesting seabirds, are relatively easy to survey, so population counts of species such as Guillemot or Puffin are probably pretty accurate. The Bittern can be censused by recording booming males and identifying each individual by its unique sound, which suggests the likely number of breeding pairs. Songbirds are much harder to survey accurately: scientists organise measured counts in a particular area over a particular period, then extrapolate from this to produce an estimate for the whole country.

Scavenging for success

The world's most numerous raptor is probably the Black Kite, which is found in tropical and warm temperate parts of Europe, Africa, Asia and Australasia. It relies on scavenging for most of its food, and has prospered thanks to our wasteful habits. In the New World the best candidate is another scavenger, the American Black Vulture, which can be found from Washington DC to Patagonia.

Why are our songbirds declining?

In recent years, both in Britain and North America, many common songbird species – such as the Song Thrush and American Redstart – have suffered major declines. The likeliest cause is habitat loss: especially as a result of intensive farming methods which reduce the amount of seed and insect food available for the birds. Other causes include climate change, which may affect birds on their wintering grounds and migration routes, and the shooting and trapping of migrant birds, which still goes on in southern Europe and other parts of the world.

Where have all the House Sparrows gone?

The decline of the familiar House Sparrow, a bird once taken for granted in Britain's towns and cities, worries a lot of people. Oddly, the fall has been patchy: sparrows have almost entirely disappeared from Central London, while in other places they are as common as before. A shortage of insects, which adults feed their young in the breeding season, may be to blame. Other factors, including modern farming methods, the shortage of suitable nest sites, and even a chemical in unleaded petrol, have also been implicated. One way to help House Sparrows is to provide 'houses' in the form of several nestboxes, which encourage this sociable little bird to breed. You can get hold of these from the RSPB.

Female House sparrow

Are Magpies and Sparrowhawks to blame?

At the same time as many of our UK songbirds have declined, we have seen a rapid increase in numbers of Magpies and Sparrowhawks, thanks to less persecution and fewer harmful chemicals. Many people have put two and two together and made five, casting the Magpie and Sparrowhawk as villains of the piece. True, both species do feed on garden birds. But they are not to blame for the songbirds' decline – indeed, many songbirds thrive alongside them. Moreover, basic ecology tells us that predator populations depend upon the fortunes of their prey, not the other way around.

Magpies are often wrongly blamed for the decline of some songbirds

No petrel shortage

The world's most numerous seabird is almost certainly Wilson's Storm-petrel, named after the Scottish ornithologist (and founding father of American ornithology) Alexander Wilson. This tiny seabird, weighing no more than a sparrow, can be found throughout the oceans of the southern hemisphere, as well as parts of the North Atlantic. Its ocean-going habits mean that it is rarely seen from land, so estimates of its total world population vary from a few million to tens of millions.

What about cats?

Unfortunately the domestic cat cannot be let off the hook so easily. Delightful though many people find them, pet cats are not a natural part of our environment. They are also very common indeed: up to 500 times as common as an equivalent wild predator such as the fox. Unfortunately, cats are nature's perfect bird-catching machines. As a result Britain's seven million or so cats kill millions of our garden birds every year.

Will global warming make things better or worse for birds?

Tricky question. In the short term, many species are likely to benefit from the effects of global warming, which so far has led to milder winters and warmer, mostly drier summers. However, as spring gets earlier and earlier this will have a negative effect on our long-distance migrants, which will soon find that their arrival times mayhave become 'out of synch' with the food supply. And if conditions in sub-Saharan Africa become drier, many migrants will also suffer on their wintering grounds, or even fail to reach them at all. There is also the outside possibility that global climate change will have the bizarre effect of 'turning off' the Gulf Stream, thus plunging north-west Europe into a new 'mini Ice Age'. This would bring incalculable devastation to our birdlife – and, for that matter, everything else.

Which birds will be affected?

If the warming trend continues, species on the southern edge of their range in Britain, such as Ptarmigan, Snow Bunting and Red-necked Phalarope, may disappear as breeding birds. Meanwhile others, such as Black Kite, Cattle Egret, Hoopoe and Bee-eater may colonise from the south. Globally, the birds most under threat

The Snow Bunting may disappear as a British breeding bird because of global warming

from global warming are those with restricted breeding and wintering ranges, when just a small change in local conditions could affect them. One example is Kirtland's Warbler, whose entire population of only a few hundred pairs breeds in the Jack Pine forests of northeast Michigan, and winters in the Bahamas. Even tiny changes in their habitat may cause breeding failure, and the species could soon become one of the first casualties of global warming.

EXTINCTION

How many birds have become extinct in recent times?

In the 400 years between the year 1600 and the turn of the millennium, just over 80 species of bird are thought to have become extinct. These include the most famous extinct creature of all, the Dodo, a large, flightless pigeon-like bird confined to Mauritius. The Dodo died out sometime in the 17th century, partly because it was hunted for food, and partly because introduced rats and cats made short work of its eggs and chicks. Either way, people were to blame.

Is extinction part of the natural cycle?

Up to a point, yes. Species whose populations go into long-term decline will eventually become extinct, and over time, new species will evolve to exploit vacant niches. But the pace of extinction is now running at a much faster rate than ever before in recorded history – thanks largely to the spread of humankind into every corner of the world, and our propensity for killing and destruction.

Has the extinction rate increased in recent years?

Yes. The extinction rate since 1600 is between 40 and 50 times what the fossil record would lead us to expect. BirdLife International recently estimated that about 1,300 species (roughly one in eight of the world's total) face the possibility of extinction during the current century. Without major changes to our attitudes and way of life this grim prediction is highly likely to come true.

The Siberian White Crane is one of the rarest birds in the world, with fewer than 4,000 individuals remaining

What made these families so vulnerable?

In the case of the rails, most were flightless, and confined to a tiny area, often on oceanic islands where they were highly vulnerable to the coming of humans and introduced predators such as rats and cats. Several of the extinct pigeons and doves also lived on tiny oceanic islands, while the parrots have mainly become extinct through habitat loss and, more recently, the capturing of birds for the cage bird trade. The moas were simply hunted to extinction for food; being flightless didn't help them.

The Dodo – an icon of extinction

Which bird family has suffered the most extinctions in historical times?

According to Errol Fuller, author of *Extinct Birds*, the rail family has suffered more extinctions since 1600 than any other family: a total of 11 species. They are closely followed by parrots (10 species), and pigeons and doves (10 species, if we include the Dodo and its close relative the Solitaire). However, if we go back slightly further in time, we find that New Zealand's 11 species of flightless moa were all driven to extinction between the 12th century, when humans first colonised the islands, and the 17th century. The demise of the moas also spelled the end for Haast's Eagle, a huge raptor weighing up to 14kg (31lbs), which preyed on the massive moas.

Hero to zero

The most numerous bird that ever existed was almost certainly the Passenger Pigeon, whose population (confined entirely to North America) may have reached ten billion, though a figure of three billion is perhaps more likely. This sociable bird travelled in vast flocks, sometimes containing many millions of birds, which blocked out the sun for hours on end. At its peak, the Passenger Pigeon population may have accounted for between one quarter and one half of all North America's birds. Yet by 1900 the species was extinct in the wild, probably as a result of persecution, hunting and habitat loss. The last captive bird, a female named Martha, died in Cincinnati Zoo on September 1st 1914. She is now on display at the US National Museum in Washington, a testament to the folly of humankind.

How do we know when a species has become extinct?

We don't. As a bird gets more and more rare, the number of sightings dwindles to a trickle, until eventually some time has passed since the last verifiable record. Often the picture is confused by unauthenticated reports, as people eager to see a rare species claim to have sighted it. Thus the last definite sighting of Eskimo Curlew in the USA was in Texas in April 1963, though another unconfirmed bird was shot in Barbados later that year. Since then there have been several claims, but none has passed the judgement of the relevant authorities. Occasionally we assume that we have found the very last wild bird alive, as in the case of Spix's Macaw, which inhabited a small area of northeastern Brazil, the only suitable habitat for the species. For several years, a lone male was seen, often tagging along with parrots of another species for company. Sadly, ornithologists visiting the area in late 2000 could not find him, despite a thorough search.

Have any 'extinct' birds been rediscovered?

Yes, several. In 1986 ornithologists in northern India stumbled across a Jerdon's Courser, a nocturnal bird that had not been seen since the early 1900s and had long been presumed extinct. Using a powerful flashlight they dazzled the bird, enabling one of the observers to walk over and

pick it up! Even more remarkably, the Four-coloured Flowerpecker was rediscovered on the Philippine island of Cebu in 1992, 87 years after it had last been seen. However, the Flowerpecker is now confined to a tiny remnant of a forest that once covered 11,000 hectares, and looks likely to become extinct during the next few years. Most amazing of all, the New Zealand Storm-petrel, last seen alive in 1850, was rediscovered in 2004, having survived undetected for more than 150 years. It is now seen regularly off the coast of New Zealand.

Which species has made the greatest comeback?

Apart from those considered extinct and then rediscovered, several species have come back from the brink of extinction, with only a handful of individuals remaining in the wild. They include Mauritius Kestrel (down to nine individuals in 1973, but now with a population in the hundreds) and the Chatham Island Black Robin. This little bird, found on an island off New Zealand, was down to five birds by 1980, with just a single breeding pair – nicknamed 'Old Blue' and 'Old Yellow'. Fortunately they still had a spark of romance in them, and managed to breed. Today's population of about 300 birds is entirely descended from those two. Less lucky was Stephens Island Wren, a probably flightless songbird from Cook Strait in New Zealand, whose population was caught and killed by feral cats.

The comeback kid – the Chatham Island Black Robin

How many species are currently at risk of extinction?

BirdLife International currently rates 1,373 species – more than one in eight of the world's birds – as being threatened with extinction, of which 198 are considered Critically Endangered. Since 2000, when they published *Threatened Birds of the World*, the number of species under threat has risen by almost 200 (about 16%).

Which continent or country has the most species at risk?

South America and Asia have the most species of bird, so it is perhaps not surprising that they also have the most at risk of extinction. Within these continents, Brazil and Indonesia are the worst cases, each with more than 100 species giving serious cause for concern. The Philippines and New Zealand high proportions of threatened species, with 15 per cent at risk of extinction. But the worst of all is Hawaii, where one-third of all native bird species are currently facing extinction.

Does Europe or North America have any species at risk of extinction?

Yes. Zino's Petrel, a nocturnal seabird confined to Madeira, is Europe's rarest breeding bird, with just 50 known breeding pairs. The most critically endangered bird that regularly migrates through Europe is the Slender-billed Curlew. It is possible that this incredibly rare wader still breeds in small numbers

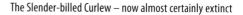
The Slender-billed Curlew – now almost certainly extinct

Whooping Cranes (top) and the California Condor (below) are two of the most endangered North American birds

somewhere in Siberia, wintering at an unknown site somewhere in North Africa or the Middle East. North America has a number of species in the 'Endangered' category, including two of the continent's largest and most famous birds, Whooping Crane and California Condor.

Going, going...

The world's rarest bird is a title with many contenders, as it is impossible to know when a particular species is down to a single individual. Until recently, the chief contender was Spix's Macaw, with a single male left in the wild until sometime in the year 2000. However, there is still a handful of individuals kept in captivity, so the bird is not yet completely extinct, though probably doomed to be so. After the last wild Spix's Macaw died, the title of the world's rarest bird passed to a Hawaiian honeycreeper named the Po'ouli. In September 2004 one of the last three wild birds was captured in the hope of starting a captive breeding programme, but on 26th November it died. The remaining two birds were last seen in February 2004, and are probably now no more.

4 · WHERE DO BIRDS LIVE?
Distribution

Are birds found everywhere?

Pretty much. Birds have conquered virtually everywhere it is possible for life to survive, apart from the ocean depths and the upper reaches of the atmosphere. They breed on every one of the world's seven continents, even in the heart of Antarctica, and can survive the most extreme habitats, from remote islands and the open ocean to landlocked deserts and barren mountaintops. In the past century or so they have even moved into our cities, to live right alongside us. Everywhere we have been – apart from outer space – birds go too.

Adélie Penguins

Where do the most birds live?

Although birds are found throughout the world, the number of species varies enormously from one region to another. Broadly speaking, there are more birds the closer you get to the Equator, with the tropics having the greatest variety and the polar regions the least. But despite having far less variety, the Arctic and the Antarctic regions still support huge populations of certain species.

What are 'faunal areas'?

Scientists have divided the world into six faunal areas (also known as zoogeographic regions): the Palearctic, covering Europe, North Africa, the Middle East and Asia north of the Himalayas; Nearctic (North America

north of the tropics); Neotropical (Central and South America); Afrotropical (sub-Saharan Africa); Oriental (South-east Asia and the Indian subcontinent); and Australasian (the area east of a boundary known as 'Weber's Line', comprising eastern Indonesia, New Guinea, Australia, and New Zealand). The world's oceans, oceanic islands and Antarctica are not included in any of these regions.

What is the 'Western Palearctic'?

The Western Palearctic is a convenient subdivision of the Palearctic region (see above), covering the whole of Europe, plus Africa north of the Sahara Desert, and the Middle East. The eastern boundary is rather arbitrary, but usually includes most of the Arabian Peninsula, Iraq (and sometimes Iran), and Russia east to the Ural Mountains. This region is the one covered by most 'European' field guides, and contains about 700 regularly occurring species, plus about 250 irregular visitors.

The Red Knot is found right across the northern hemisphere

What is the 'Holarctic'?

'Holarctic' is the name given to the Nearctic and Palearctic regions combined. A number of circumpolar species, such as Red Knot, have a Holarctic breeding range.

Packing them in

The country with the highest number of recorded species is either Colombia or Brazil, both of which are home to around 1900 species. They are closely followed by Peru (1880), Ecuador (1660), and Indonesia (roughly 1600). Six of the top ten most species-rich countries are in South America. In comparison, just over 900 species have been recorded in North America north of the Mexican border, and a shade under 600 species in the United Kingdom.

Which regions have the most species?

The most bird-rich regions – at least in terms of the number of species – are the zones that cover the tropical and equatorial areas of the world, i.e. the Neotropical, Afrotropical and Oriental. The Neotropical region contains about 3,000 species (roughly 30 per cent of the world's total), while the Palearctic and Nearctic regions have only about 1,000 and 750 species respectively. Despite its smaller size, the Australasian region boasts about 1,600 species, many of which are only found in that part of the world.

Which habitat supports the greatest variety of species?

Undoubtedly the tropical rainforest, which supports a greater variety of all animal life than any other habitat. Found across parts of South America, equatorial Africa and southeast Asia, this is also one of the world's most threatened habitats, due to pressures from human exploitation. As a result, many rainforest species are at risk of extinction.

Which habitat supports the smallest variety of species?

The two polar areas, especially Antarctica, which has only a handful of regularly occurring breeding species, most of them penguins. However, many more species breed in the sub-Antarctic islands such as the Falklands, South Georgia and Snares Island, off the southern coast of New Zealand.

The Latin Quarter

The most bird-rich continent is undoubtedly South America, which boasts at least 3400 species, almost one in three of all the world's bird species.

The diversity of species in the most bird-rich continent, South America, is truly incredible

Continental thrift

The least bird-rich continent is Antarctica, which despite being roughly twice the size of Australia, has only 45 or so species, only around a dozen of which breed there. However, many other seabird species can be found in the oceans around the continent.

What is a 'montane' species?

Clark's Nutcracker is a montane species found in North America

This term applies to a species which normally lives and breeds in mountainous areas. In Britain this includes birds such as the Ptarmigan, Snow Bunting and Dotterel, even though in other parts of Europe the latter two species may breed at sea level. In North America birds such as Clark's Nutcracker and Mountain Chickadee are considered to be montane species, generally living at above 3,000 metres (over 9,800 feet). Montane species are usually altitudinal migrants, moving down to lower altitudes in winter to find food.

What is a 'pelagic' species?

This refers to ocean-going seabirds, which spend most of their lives at sea, usually coming to land only to breed or when driven ashore by very bad weather. Pelagic species include auks, gannets and boobies, frigatebirds, and to a lesser extent gulls, terns, cormorants, phalaropes, divers and penguins. But the birds best suited to the oceanic lifestyle are undoubtedly the tubenoses – the 120 or so species of albatrosses, petrels, storm-petrels and shearwaters that are perfectly adapted to living for long periods at sea. The 'tubenose' is a protuberance on top of their bill that enables them to smell food from many miles away, and to excrete excess salt from drinking seawater.

Frigatebirds are known as pelagic species as they spend much of their life at sea

What is a 'cosmopolitan' species?

It's any species that breeds over much of the globe, including both the 'Old' and 'New' Worlds, as opposed to most others, which are confined to just one or two of the world's regions. Examples include Osprey, Peregrine, Barn Owl, Cattle Egret and – by virtue of extensive introductions – House Sparrow.

Azure-winged Magpies are found in Iberia and China, but nowhere in between

The Cattle Egret is one of the most widespread bird species in the world, and has been seen on all seven continents

What is a 'relict' species?

One that was once found over a wide geographical area, but is now confined to a relatively small part of its former range. Sometimes two populations have ended up isolated from one another, perhaps thousands of miles apart. The Azure-winged Magpie is found in China and Japan, and has a completely separate population in Spain and Portugal. It used to be thought that explorers had brought this species back to Western Europe during the 15th and 16th centuries. However, fossil evidence now suggests that it once occurred naturally right across Europe and Asia, but has disappeared from all but the westernmost and easternmost extremities of its former range. Another example is the Caspian Tern, whose range extends over much of the Old and New Worlds, yet is highly fragmented, with scattered colonies located many hundreds of miles apart.

Where the rare things are

The country with the most species with a restricted range (defined by BirdLife International as being less than 50,000 sq. km / 19,300 sq. miles) is Indonesia, with over 400 such species.

What is an 'endemic' species?

An endemic species is one confined to a restricted geographical area. This area could be a faunal region, a country, or even just a single island, depending on how you wish to define it. An endemic may thus be very widespread or very localised. So despite being the world's commonest bird, and found across much of sub-Saharan Africa, the Red-billed Quelea can still be categorised as an African endemic. At the other extreme, the Razo Lark is confined to a tiny island in the Cape Verde archipelago, with fewer than 100 individuals in an area of less than seven square kilometres (2.7 square miles).

The Red-billed Quelea, a species of finch confined to sub-Saharan Africa, is nevertheless the most abundant bird in the world

Bird-free zones

The country with fewest species is the Pacific Island of Nauru, with 27 (11 of which are waders), including a single endemic, the Nauru Reed Warbler. Easter Island, also in the Pacific Ocean, has only nine breeding species, but is not a sovereign state so cannot strictly be counted as a 'country'. Both islands were devastated at the hand of man – Nauru by phosphate mining and Easter Island by deforestation – so it is not surprising they are so poor for birds. If breeding species alone are counted, Vatican City probably has the fewest, though if 'flyovers' are counted, its national list is undoubtedly longer than either of the Pacific islands.

Which region has the most endemic families?

The one with the most endemic families is, not surprisingly, the Neotropical region, with 31, including rheas, tinamous and toucans. Its nearest rival, the Australasian region, has 16 families found nowhere else in the world, including Emu, Plains-wanderer and the scrub-birds. By contrast, the Palearctic has just one endemic family (the accentors), while the Nearctic has none.

The rheas are endemic to South America

Opposite: The island of Madagascar, off the east coast of Africa, is home to many endemic species

Why do endemic species occur?

Most endemic species are found in specific hotspots, known as 'areas of endemism'. These include oceanic islands, mountainous regions, and some lowland habitats such as tropical forests. The common factor is that at some time in the distant past their birdlife became isolated from other, similar areas – either by the sea, or by natural barriers of unsuitable habitat such as grassland. As a result, isolated populations eventually evolved into different species from their counterparts elsewhere. Endemism in birds often goes hand in hand with that in other groups of plants and animals – as in Madagascar, where a largely unique fauna and flora has evolved, including the famous lemurs, as well as endemic bird families such as the vangas and mesites.

Why do some islands have lots of endemics, while others have very few?

Islands that have been isolated for several million years tend to have more endemics than those which until recently were connected to another land-mass. So Jamaica has about 27 endemic species, including four hummingbirds, two parrots and a wood-warbler, while nearby Trinidad has only one, the Trinidad Piping-guan. This is because Jamaica has existed independently for millions of years, while Trinidad was connected to the South American continent until only a few thousand years ago.

Do some birds' ranges change over time?

The Collared Dove was unknown in Britain until the 1950s, but is now a common and widespread breeding bird

Yes. Just like rises and falls in bird populations, changes in range frequently occur, and for all sorts of different reasons. Expansions, when a species colonises a new area, may arise from climate change, a new food source, or human intervention – either deliberate or accidental. Likewise, contractions may result from many factors, including persecution, loss of habitat and climate change.

How far can a species expand its range?

There's no stopping certain species. If the conditions are right, some even manage to reach new continents. The classic case is the Collared Dove, which during the 20th century spread north and west from its original breeding range in south-west Asia to colonise virtually the whole of Europe, and has now gained a foothold in the New World too – though it may have had a helping hand to get there. The original impulse to explore new areas was probably the result of a genetic mutation, but the reason for the birds' success was finding a vacant ecological niche, which they were able to exploit. The Cattle Egret is the best known that has spread naturally from the Old to the New World in historical times, having reached South America from West Africa following a storm in the early 20th century, and since spread northwards to become a common sight in parts of North America.

The Ivory Gull is one of the few species that breeds in the High Arctic

On top of the world

The title of the most northerly bird is shared between three species: Black-legged Kittiwake, Snow Bunting and Northern Fulmar, each of which has been seen at the North Pole itself. The most northerly breeding bird is Ivory Gull, which has been found nesting on the edge of the Arctic pack ice at 85 degrees North.

Do human beings influence changes in range?

As we continue to rearrange the face of the earth to suit our own ends, we exert a disproportionate influence over the populations and distribution of birds. This is not a recent phenomenon: Neolithic man cleared forests in order to plant crops, thus presumably benefiting grassland species at the expense of forest ones. But during the second half of the 20th century, the effects on the environment began to have a far greater impact. Modern agriculture, road and house building, pollution and persecution have all eaten deeply into the range of many species.

Are some birds more adaptable than others?

Yes. Many birds have learned to live alongside human beings, and some resourceful species have positively flourished in our company. For example, gulls have long since ceased to live

Chilled out

The most southerly bird is the aptly named South Polar Skua, which has been sighted at the Russian base at Vostok in Antarctica. This is officially the coldest place in the world, where temperatures have dropped as low as minus 89.2°C. However, the world's hardiest bird is surely the Emperor Penguin. Not only does it spend its whole life on the Antarctic ice, but it even chooses to breed in winter, when temperatures can drop lower than minus 45°C. No wonder polar explorer Apsley Cherry-Garrard described his quest to find breeding Emperors as "the worst journey in the world". The most southerly breeding bird is, however, not the Emperor Penguin, but the Antarctic Petrel, which has been recorded nesting at 80°30′ South.

A Kestrel hovering overhead as it searches for food

exclusively on the coast, and now nest on high-rise buildings in cities, where they find rich pickings among what we leave behind. Other examples include Kestrels hunting for voles along motorway verges throughout Britain, Peregrines and Red-tailed Hawks nesting on skyscrapers in New York's Manhattan, and various kinds of parakeets, which thrive in city parks from London to Miami.

How can we be sure about distribution changes?

As with changes in population, we discover the detail about range changes through surveys, many of them carried out by amateur birdwatchers. The very first nationwide *Atlas* survey of breeding birds was carried out in Britain and Ireland from 1968–1972, and involved more than

The Short-toed Eagle has a broad range across Europe, Africa and Asia, yet is a monotypic species with only one race

10,000 amateur fieldworkers. These dedicated observers surveyed every single one of almost 4,000 10-kilometre squares, producing more than a quarter of a million individual records. Since then, similar surveys have been carried out in many other parts of the world, not just on breeding birds, but on winter visitors and migrants too.

What is a 'monotypic' species?

One with a single race across its entire range. One example is the Short-toed Eagle, which breeds in Europe, north-west Africa and Asia, but whose populations do not vary enough from one another to be separated as different subspecies.

85

Do species vary across their geographical range?

Yes, sometimes. Species where one population shows consistently distinct characteristics from another population elsewhere in its range are known as 'polytypic', with two or more different races being recognised (see Chapter 2). Examples include the Eurasian Wren, which has at least 13 recognised races in its range across Europe and Asia. These may vary considerably in size, colour and overall appearance, though they are all recognisably the same species. Many races occur on remote offshore islands, of which the most famous is the 'St. Kilda Wren', which is slightly larger, darker and has a noticeably longer bill than its counterpart on the Scottish mainland. Other races of Wrens occur on Shetland and even on the tiny island of Fair Isle.

The Eurasian Wren (left) and St Kilda Wren (right) are two races of this familiar species

Grey Partridges are native to Britain, unlike their cousin the Red-legged Partridge, which was introduced here

What is a 'cline'?

Not all birds can be neatly classified into distinctive subspecies: many vary gradually from one end of their range to another, for example getting smaller or larger, darker or lighter. This phenomenon is known as a cline, and occurs most frequently in colonial seabirds.

What does 'indigenous' mean?

An indigenous species is simply one that has evolved naturally to live in a particular area, without having been introduced there by humans. In other words, a native. If a species expands its range as an indirect result of human activity, such as the clearing of forests, it is still considered indigenous to that area. But if it has been deliberately or accidentally introduced by humans, then it is not considered indigenous. For example the Pheasant and Red-legged Partridge are non-native species in Britain, having been introduced for the purpose of shooting, while the much rarer Grey Partridge is a native species. Conversely, in North America the Grey Partridge is a non-native species, having been brought by humans across the Atlantic Ocean from Europe.

A bird's world

Many species vie for the title of world's most widespread wild bird.

The Cattle Egret colonised Australia from Asia and North America from Africa (via South America). Wilson's Storm-petrel is found over most of the world's southern oceans, though is less common north of the Equator. Several waders, notably Sanderling and Ruddy Turnstone, are found on most of the world's coasts, from the Americas and Europe to Africa, Asia and Australia.

The world's most widespread passerine is the Barn Swallow, which occurs on every continent apart from Antarctica. The Horned (or Shore) Lark also has a vast breeding range, spanning North America, Europe, northwest Africa and Asia, plus a tiny population in the Andes.

The Barn Swallow (top) and Horned (or Shore) Lark (bottom) are both widespread species globally

What is a 'naturalised' species?

One which, having been introduced by humans (either deliberately or not) has now established a fully self-sustaining population. The distinction between 'naturalised', 'introduced' and 'feral' is not always clear. Many people use them interchangeably, although strictly speaking only 'introduced' implies a deliberate release.

Does 'alien' mean the same as 'introduced' or 'naturalised'?

In practice, yes. However, the word 'alien' has negative cultural associations, due in part to its link with invaders from outer space, and more recently its use to describe illegal immigrants. It is also associated with more destructive introduced species such as the Ruddy Duck, Mink or Grey Squirrel, which were brought from their native North America to Britain and Europe, and the House Sparrow and Starling, which were taken in the

The Ruddy Duck is a non-native species in Europe

opposite direction from Britain and Europe to North America. Nowadays conservationists try to avoid the word, preferring more neutral alternatives such as 'exotic' or 'non-native'.

What is a 'feral' species?

A 'feral' species is one that is now living freely in a wild state, having either escaped or been released from captivity. It is often regarded as a kind of 'halfway house' between 'tame' and 'wild', and ceases to apply when a population has become entirely self-sustaining. So the Mandarin Duck is no longer regarded as feral in Britain, whereas its close relative the Wood Duck is still considered so, because it has yet to establish a fully wild population. However, the best-known species in this category, the Feral Pigeon, retains the title despite being one of our commonest and most successful birds.

What is a 'reintroduced' species?

A reintroduced species is one which was once indigenous to a particular area or country, then died out, but has since been returned to the wild by humans. British examples include the Capercaillie, reintroduced into Scotland in the 19th century after disappearing in the 18th, and the White-tailed Eagle, which previously last bred in Britain in 1916, disappearing due to persecution by hunters and gamekeepers, but is now back in northern Scotland and thriving.

The Capercaillie was reintroduced to Britain in the nineteenth century

Birds with altitude

The highest bird recorded on land is the Alpine (or Yellow-billed) Chough, a small group of which followed a climbing expedition to a height of 8,150 metres (26,500 feet) on Mount Everest in 1924. They survived by foraging on scraps of the climbers' food as they went. The Alpine Chough has also been recorded nesting in the Himalayas at altitudes of up to 6,000 metres (19,500 feet) above sea level. The lowest altitude at which birds have been recorded is the Dead Sea depression in Israel, where Little Green Bee-eaters have been found nesting at almost 400 metres (1,300 feet) below sea level.

Why are birds introduced into a new area?

The exotic Lady Amherst's Pheasant was introduced to Britain from Asia

Introductions are not always deliberate: exotic cage birds often escape, and while most fail to survive in their hostile new environment, a few thrive and establish self-sustaining breeding populations. Otherwise, the most common reason for introducing birds to a new area is to hunt them for food. This accounts for the presence of several exotic gamebirds in Britain, including the Pheasant, a southwest Asian species originally brought by the Romans. Others were introduced purely for their aesthetic value, including Golden and Lady Amherst's Pheasants (from Asia) and Little Owl (from continental Europe). In North America, the Starling was released in Central Park in the late 19th century by a misguided group of philanthropists who believed every bird mentioned in Shakespeare should be found in the USA. Today, Starlings can be found from Alaska to Baja California, and almost everywhere in between. They have also been introduced to New Zealand and South Africa.

Starlings are native to Europe, but were introduced to North America in the nineteenth century

Which introduced birds are most likely to do well?

Parrots have a strong track-record in this respect – especially the smaller parakeets, whose adaptable behaviour and gregarious habits enable them to gain a rapid foothold. Today, only a few decades after their original release, Rose-ringed (or Ring-necked) Parakeets are a common sight in London suburbs, with one roost containing several thousand birds. Originally from northern India, this species has also established feral populations in many other parts of the world, including West Africa and the USA. Monk Parakeets, from South America, have thriving colonies in parts of the southern United States, and several European cities including Barcelona. Crows are also highly adaptable: the Indian House Crow has hitched a ride on ships to many places, including the Netherlands, the Red Sea resort of Eilat in Israel and even Durban, South Africa.

Which introduced birds have done the most harm?

Compared to such destructive mammal pests as the Grey Squirrel, Mink and Coypu, birds may seem relatively harmless, but they can still harm native wildlife and ecosystems. Because introduced species have few, if any natural predators, they are often able to expand their range and numbers until they reach pest proportions. In Britain, Canada Goose and Ruddy Duck have both thrived in the absence of competition, and the latter species then travelled to mainland Europe where it started to hybridise with a threatened native species, the White-headed Duck, leading to a scheme to eradicate the species from western Europe. Elsewhere in the world, adaptable species such as the House Sparrow and Starling compete for food and nest-sites with less adaptable native species; while unique ecosystems such as those found on oceanic islands can be devastated by deliberate or accidental introductions.

The Rose-ringed Parakeet is now a familiar sight in London parks, having been introduced from India in the late 1960s

World beater

The most successful introduced bird is probably the humble House Sparrow, which was originally found only in Europe and parts of Asia and North Africa. House Sparrows have been successfully (or disastrously, depending on your point of view) introduced to North America, South Africa, New Zealand, and many island groups – including the Falklands and Hawaii. Today the House Sparrow occupies an area of land equivalent to about one-quarter of the earth's land surface.

Which part of the world has had the most introductions?

The two places which have had the most species introduced into their avifauna are Hawaii, with more than 160 species, and New Zealand, with more than 130 – including such familiar British birds as the Song Thrush, Blackbird and Goldfinch. However, the Miami area of Florida probably has more exotic species living in a 'semi-wild' state than any comparable area, as escaped tropical parrots and other birds thrive on the area's warm climate and plentiful supply of food.

Which is the most bizarre introduction?

There are several candidates, but surely the most incongruous was the presence of a small colony of the Greater Bird-of-paradise, originally from New Guinea, on the island of Little Tobago in the Caribbean. Originally introduced there by a wealthy English newspaper proprietor in 1909, the birds managed to survive until a major hurricane wiped out much of their habitat in 1963, which led to an irreversible decline. The last bird was seen in 1981.

Home made

The country with the most endemic species is Australia, with well over 300 endemics. However, the island of New Guinea (which comprises two states: Papua New Guinea and Irian Jaya) has about 320 endemics. The country with the highest proportion of endemics is Madagascar, where endemics account for more than one in three of all bird species. More than 130 countries have no endemics at all. The countries with the lowest proportion of endemics are Guyana and Ghana, with none, despite each having more than 700 species on their national lists.

5 • HOW DO BIRDS MOVE?
Locomotion

IN THE AIR
Are birds the only animals that can fly?

Obviously not, since millions of insects are also pretty good at it. But birds reign supreme among vertebrates (animals with backbones – also including mammals, reptiles, amphibians and fish). In fact, the only other vertebrates capable of powered flight (as distinct from gliding, as practised by certain 'flying' squirrels, fish, frogs, lizards and snakes) are bats. And bats have evolved nothing like the same variety of forms and techniques. Just think of the completely different ways in which, say, eagles, hummingbirds, albatrosses, ducks and larks take to the sky, and you will begin to appreciate birds' amazing mastery of the air. As the American bird artist Roger Tory Peterson put it, "Birds have wings – they travel."

Flying Fox

What make birds so good at flying?

Apart from wings and feathers (see Chapter 1), their other important adaptations include thin, hollow bones and a lack of heavy teeth and jaws, to help reduce their body weight. They also have a highly efficient blood circulation and respiratory system, and a high metabolic rate, which increase their ability to convert energy into flying power.

How did flight evolve?

In birds, the ability to fly is inextricably linked with the evolution of feathers. Their lightness, strength and aerodynamic shape were the key to getting birds into the air. It seems most likely that the earliest birds first developed feathers on their wings and tail, which enabled them to glide for short periods. As the development of flight feathers progressed, so the descendants of these ancestral birds were able to fly further, higher and for longer periods. Natural selection gave the best flyers the highest chances of survival. So over many, many generations – covering hundreds of thousands, perhaps millions, of years – flight evolved. It eventually became their standard means of getting around.

How do birds get airborne?

The first challenge faced by birds (and indeed by any flying object, from aphid to Airbus) is getting airborne. To do this, a bird has to overcome the force of gravity by generating enough energy to lift itself into the air and stay there. Small birds are light enough simply to flap their wings and take off, but larger ones such as ducks and geese may have to build up forward motion by running and flapping across the surface of the water. Really heavy birds, such as bustards and swans, must put in even more effort over an even greater distance in order to become airborne.

The sky's the limit

The world's most airborne bird is probably the Sooty Tern, which may not land for an amazing seven or more years after leaving the colony where it was hatched. The Common Swift is also highly adapted to an aerial existence, with some individuals not touching down for two to three years between fledging and first breeding, during which time they will have migrated to Africa and back – twice! Both these birds sleep and feed on the wing, the swift snapping up tiny flying invertebrates, and the tern snatching morsels of food from the waves.

Swifts are amongst the most aerial of all birds

How do birds stay airborne?

Once up, a bird's next challenge is to stay up. To do so it must overcome the problem of drag that results from its own forward motion. It does this by generating enough lift to overcome drag and reach an equilibrium. Different species use very different methods to stay in the air: some generate lift by flapping their wings rapidly; others have long and/or broad wings for gliding; and heavier species such as raptors soar on broad wings, taking advantage of rising air currents/thermals (see page 100).

How long can a bird stay airborne?

In some cases, almost as long as it wants. Several birds, including swifts, albatrosses and the Sooty Tern, are known to spend almost unimaginably long periods on the wing without ever landing; indeed they only touch down eventually because no bird can lay eggs and raise young in mid-air. Swifts can feed, sleep and even copulate while in flight; testament to the incredible adaptability of birds (and a glimpse of what pleasures we earth-bound mammals might be missing).

How high do birds fly?

Anything from just above sea level to an altitude of several thousand metres – depending on the species, and whether it is travelling a short distance or undertaking a long migratory journey. Most birds seldom have any reason to rise above 500 metres (1,600 feet) or so. But extra-efficient respiration, plus the ability to survive extremely cold temperatures, allows some to reach extraordinary altitudes if required. Vultures, swans and geese have been recorded flying at jet airline heights, in conditions that no mammal could possibly survive.

The frigatebirds have the lowest wing-loading of any bird species

What is 'wing-loading'?

This technical term takes us into the territory of flight engineers. Essentially, it is the ratio between the wing area and the weight of the bird; thus, the greater a bird's wing-area in relation to its weight, the lower its wing-loading. Birds with a low wing-loading, such as vultures and condors, are able to soar for long periods without flapping their wings and using up energy; whereas birds with a high wing-loading, such as swans and geese, must flap their wings constantly. The group of birds with the lowest wing-loading of all is the frigatebirds, which can hang in the air almost effortlessly for hours or days on end. In fact, so light are these enormous seabirds, that their skeleton actually weighs less than their feathers.

How many different flight techniques do birds use?

Red Kites are expert flyers, able to soar and glide to save energy when airborne

As well as the standard means of flapping their wings, birds have developed several more specialised techniques to stay airborne or reduce the expenditure of energy. These include gliding – using outstretched wings in order to travel forward without flapping (e.g. Sparrowhawk); soaring – similar to gliding but generally in a circling pattern, using air currents to gain or maintain lift (e.g. Buzzard); and hovering – flapping wings very rapidly in order to stay in one place for a short period of time (e.g. Kestrel). Basic wing shape means that every species tends to favour a particular technique. But wings are adjustable, and most birds will adapt to suit the conditions. For example, Sparrowhawks sometimes soar, while Buzzards often glide, and sometimes even hover – albeit more clumsily than Kestrels.

Why do some species glide, soar or hover?

Gliding and soaring are mainly practised by larger, heavier birds such as eagles, hawks and gulls, so they can stay airborne longer (an advantage when searching for food over a large area). Seabirds such as albatrosses and shearwaters ('stiff-wings') practise 'dynamic soaring'. This is a technique that enables them to glide great distances over the ocean's surface without flapping, even in windless conditions, by using the updraft from wave slopes. Both soaring and gliding conserve energy. Hovering, by contrast, tends to use a lot of energy for a very brief period. However, it allows a bird to remain stationary in mid-air, enabling, for example, a Kestrel to spot prey in the verge below or a hummingbird to sip nectar from flowers.

Do birds ever fly backwards?

The only group of birds whose members regularly fly backwards is the hummingbirds. They have evolved a special flight technique, involving flexible shoulder-joints and extremely fast wing beats, which enables them to move in any direction – forwards, sideways, up, down or backwards – in order to position themselves perfectly for sipping nectar from a hanging flower. Other birds occasionally fly backwards for very brief periods, either accidentally, when caught by the wind, or deliberately, in order to avoid attack or catch their prey.

Can birds fly upside-down?

No birds habitually fly upside-down, though again some birds may do so for very brief periods by accident. A few will also flip or roll over during aerial courtship displays, including various raptors and – of course – rollers.

Ruby-throated Hummingbirds can, like other members of their family, fly backwards

How fast do birds usually fly?

Flying speeds vary considerably, depending on the group or species involved, and whether it is on a short flight or a long migratory journey. Most migrating songbirds fly at speeds of between 25 and 40km/h (15–25mph), which is the optimum range for balancing the expenditure of energy and the distance covered. Larger birds often fly faster: cruising speeds for shearwaters and albatrosses are about 40–55km/h (25–35mph), while ducks, geese and swans can maintain speeds of up to 70–75km/h (44–47mph) over long distances.

Birds such as terns occasionally fly upside down for short distances

What are thermals?

Thermals are rising columns of warm air, produced by the radiant heat of the sun on the surface of the earth. They generally occur from mid-morning onwards, as the sun rapidly raises the temperature of the cooler ground. Large, heavy birds such as vultures, hawks and buzzards use thermals in order to gain height as quickly as possible without expending valuable energy by flapping. Once they have

Raptors such as Turkey Vultures use thermal air currents to gain altitude

reached their desired altitude, they will then glide away towards a new feeding area. Thermals are also important for many raptors and other large birds during migration, especially when faced with a wide expanse of water. Birds crossing the Straits of Gibraltar or the Red Sea will use thermals to gain height, before gliding across the water to the other side.

How does the weather affect flying birds?

In all sorts of ways: some a help, others a hindrance, and a few downright fatal. As well as using thermals to gain height, or air currents over the open ocean for 'dynamic soaring' (see page 98), many migratory birds use particular weather conditions to aid them in their journey. For example, light following winds produced by high-pressure systems in spring help carry northward-bound migrants to their destination more quickly.

Stormy weather can cause problems for migrating birds

Speed freaks

The world's fastest flying bird is probably the Peregrine Falcon, whose speed during its 'stooping' dive has been measured at around 400 km/h (250 mph). The White-throated Needletail (a type of swift from Asia) has been recorded at 170km/h (106mph). In level flight, the Common Swift has recently been found to reach a top speed of 111.6 km/h (69.3 mph), while flying both horizontally and upwards – so swifts really do live up to their name!

How does bad weather affect migrants?

Migrating birds are at the mercy of the elements. Unusually windy or stormy weather may drive many off their intended course, sometimes with unexpected results. Every autumn, millions of North American landbirds head out into the Atlantic Ocean on their journey south, but strong winds – especially in the aftermath of a hurricane – may push them eastwards. The vast majority of these displaced birds will become exhausted and perish in the waves, but a few will cross the Atlantic and make landfall in Britain or Ireland.

Why are some birds flightless?

Given that flying is virtually synonymous with birds, it does seem odd that a few cannot do it at all. But in fact flightlessness can make perfect sense, so long as it confers benefits that will give the individual an advantage in the evolutionary 'arms race'.

What are the advantages of flightlessness?

That depends on the environment in which a particular bird lives. Flightless species fall into several categories: some are large, terrestrial birds such as the Ostrich, for which sheer size is a sufficient deterrent to most land-based predators. If these birds had retained the power of flight, they would never have been able to evolve to such gargantuan proportions (the Ostrich is, for example, about ten times the weight of the largest flying birds). Flightless species also thrive in aquatic environments, especially the ocean: in the case of penguins, modified wings have become flippers, enabling them to dive to great depths.

What are the *disadvantages* of flightlessness?

Today, the most serious one is not being able to fly away from trouble! But this was not a problem until human beings began to explore the world – otherwise flightlessness could not have evolved in the first place. Many flightless species arose on oceanic islands, where a dearth of predators

Penguins almost appear to 'fly' underwater

Mile-high club

The highest-flying bird was a Rüppell's Griffon Vulture, which was sucked into a jet engine over the Ivory Coast in West Africa, at an altitude of 11,300 metres (more than 36,700 feet), in November 1973. The greatest height at which living birds have actually been observed was 9,375 metres (roughly 30,500 feet): a flock of Bar-headed Geese flying over Mount Everest. In Britain, a pilot spotted a flock of Whooper Swans flying over the Hebrides at a height of around 8,200 metres (26,650 feet) in December 1967.

made flying unnecessary. However, as soon as people arrived, their chances of survival dropped dramatically. This was partly because they were easy to catch for food, and partly because humans introduced predators such as rats, cats and dogs, which found them easy prey. Two of the best-known examples of flightless birds that were casualties of human expansion are the Great Auk and that proverbial icon of extinction: the Dodo.

Did flightless birds have flightless ancestors?

Odd as it may seem, all flightless birds evolved from flying ancestors. They did so through a process known as 'neoteny', or arrested development, in which characteristics normally found in the embryonic state of a bird persist into adult life. Thus all flightless birds have wings of some sort, however vestigial they may appear. And

The male Ostrich has a spectacular courtship display

and rats. Errol Fuller (author of *Extinct Birds*) considers that up to one quarter of the 80 or so species to have become extinct since 1600 were flightless. These include several species of rail confined to oceanic islands, and massive flightless giants such as the moas of New Zealand and the *Aepyornis*, or elephant birds, of Madagascar.

Does flightlessness run in families?

Yes, with a few exceptions. The best-known family of flightless birds is the penguins, all of whose 17 species are flightless – though if you see a penguin swimming underwater you might argue that it has not lost the power of flight, but simply adapted it to a different medium. The 13 currently recognised species of ratite (Emu, Ostrich, rheas, cassowaries and kiwis) comprise the other group whose members are all flightless. Flightlessness is also prevalent amongst rails: almost one in four living or recently extinct species are, or were, flightless. Other groups with flightless members include the grebes, cormorants, ducks, parrots and the unique Kagu of New Caledonia – a peculiar species distantly related to the rails. The only passerine that has ever been described as flightless (and even then, it may have retained limited powers of flight) was the now-extinct Stephens Island Wren, all known specimens of which were discovered (and unfortunately killed) by feral cats.

these wings still have a job to do: in some cases they have evolved into very different structures, such as the flippers of penguins, which propel them through the water; in others, including the Ostrich, they are used in courtship display.

How many of the world's birds are flightless?

Fewer than 50 species, or less than 0.5 per cent. However, many other flightless species have become extinct due to the invasion of their environment by pests such as cats

IN THE WATER

What proportion of the world's species regularly swim?

Of the world's 10,400 or so species of bird, only about 400 (less than four per cent) habitually swim on or dive under water. Those that do are mostly found in particular groups: divers, grebes, tubenoses (albatrosses, shearwaters and petrels), pelicans, cormorants, wildfowl (ducks, geese and swans), coots and gallinules, phalaropes, gulls, skuas, terns and auks. Several other groups, such as herons and egrets, flamingos and waders, can also be classified as 'waterbirds', and these include some species that do occasionally swim for short periods.

How have swimming birds adapted to an aquatic existence?

In several ingenious ways. For example, many have webbed feet to increase propulsion when swimming or diving. Most have fully webbed feet, while other groups such as grebes and phalaropes have either partial webs, or lobes on the sides of the toes. Water is a harder medium than air through which to make headway, so some aquatic species such as penguins have developed much tighter feather patterns in order to create a more streamlined profile, making their plumage look more like mammalian fur than feathers. Penguins also have dense bones, unlike other birds whose bones are mostly hollow and filled with air, helping them to submerge and dive more easily.

Dalmatian Pelicans are well adapted to an aquatic existence

Flipper power

The world's fastest swimming birds are penguins, which can swim at speeds of about 15 km/h (9.3mph) in pursuit of their prey. Gentoo Penguins have been timed at over 35 km/h (22mph) over short bursts – roughly five times as fast as an Olympic freestyle swimming champion. Penguins are also undoubtedly the most aquatic birds. Some species may spend as much as three-quarters of their daily lives in the water, and outside the breeding season will stay away from land completely for several months at a time.

Why don't waterbirds become waterlogged?

Many birds that habitually swim or dive have specially adapted oil glands at the base of their tail, from which they secrete an oily substance that they use to waterproof their feathers. These glands are particularly well developed in groups that spend much of their time in the water, such as penguins, ducks and gulls. However, a few waterbirds such as the frigatebirds have very small oil glands and so cannot coat their plumage adequately. These birds avoid landing on the sea, otherwise they would rapidly become waterlogged and drown.

Cormorants are expert at diving underwater to catch fish

Do all birds dive in the same way?

No. Most, including penguins, divers, grebes, cormorants, auks and several species of ducks, dive from the surface of the water. Some of these, such as penguins and auks, propel themselves using their wings; others, like divers, grebes and ducks, use their powerful webbed or lobed feet. A few birds, including gannets, boobies, terns, kingfishers and fish-eating raptors, dive into the water from the air. Some, such as gannets, plunge in deeply; others, such as fish eagles, prefer to snatch prey from the surface.

Are there any aquatic songbirds?

Yes: the five members of the dipper family. These extraordinary little birds have adapted to feed underwater, and even walk along the bottom of streams and rivers – a skill shared by very few other warm-blooded creatures except the Hippopotamus! Their adaptations include dense plumage, flaps over their nostrils and a specially adapted transparent eyelid enabling them to see their prey while keeping the water out.

Plumbing the depths

The deepest diving bird is the Emperor Penguin, which has been reliably recorded at a depth of 265 metres (875 feet), though a depth of more than 565 metres (1,865 feet) has been claimed. Its smaller relative the King Penguin has been recorded at depths of 240 metres (790 feet). Amongst flying birds, both Guillemots and Brünnich's Guillemots (Common and Thick-billed Murres) have been recorded at depths of up to 200 metres (660 feet). The longest time a bird can stay underwater is about 27 minutes, a record also held by the Emperor Penguin. However, cetaceans can easily outdo any bird when it comes to diving: Sperm Whales have been recorded 2,500 metres (1.6 miles) below the surface, and can stay underwater for over an hour.

ON LAND

How do birds get about on land?

Some run or walk. Others hop, using both legs together in a single motion. By and large, non-passerines do the former, while – with a few exceptions, including crows, starlings, pipits and wagtails – passerines prefer the latter.

Hummingbirds have tiny feet and so are unable to perch on the ground

Which birds move best on the ground?

Many aquatic birds such as grebes and divers are pretty useless on terra firma; and aerial species such as albatrosses, shearwaters and many birds of prey tend to move very clumsily. Some species, however, have become expert runners. These include the larger ratites (ostriches, rheas etc.), game birds such as partridges, waders like the Sanderling, and of course the two species whose name celebrates their prowess: the roadrunners.

Roadrunners, as their name suggests, are expert at running fast

Which birds are the least capable of walking?

Swifts and hummingbirds, whose legs and feet are very small indeed, are pretty much helpless on the ground. In fact most swifts can do no more than cling to vertical surfaces with their forward-pointing toes (the name of their order, Apodiformes, means 'lacking feet'). Also, divers and grebes have legs set so far back on their body that they find it very difficult to propel themselves while on land. For swifts and hummingbirds the inability to walk

Sprint finish

On the ground, the world's fastest running bird is the Ostrich, which typically runs at around 45–50km/h (28–39mph) and can reach speeds of more than 70km/h (44mph) over short distances – faster than most birds can fly. The fastest mammal, the Cheetah, can sprint at more than 100km/h (62mph) over short distances. The fastest runner amongst flying birds is the Greater Roadrunner, which far prefers running to flying, and can reach speeds of up to 40km/h (25mph).

is the trade-off for an essentially aerial existence; for divers and grebes it is the trade-off for an aquatic one.

How do woodpeckers climb trees?

Woodpeckers (and other arboreal groups such as creepers and nuthatches) have special adaptations for climbing up the trunks and branches of trees. In the case of woodpeckers, most species have two toes pointed forward and two pointing back, enabling them to grip the surface of bark. Woodpeckers also have a stiff tail, which they use as a prop to stabilise themselves – an adaptation shared by birds such as the Treecreeper, which also has long claws for clinging to the bark.

Woodpeckers are specially adapted to climb vertical surfaces of tree trunks

from the way they place their feet when climbing: one foot above, the other below; enabling them to move up, across or down the surface of a tree.

How do nuthatches climb down trees?

The world's 25 or so species of nuthatch share one characteristic unique amongst birds: the ability to climb down headfirst, as well as up, trees. This is despite the fact that they lack the stiff tail of woodpeckers and creepers. Instead, their ability derives

Nuthatches are the only family of birds that can climb down tree trunks as well as up

ALL TOGETHER

Why do birds form flocks?

For a number of reasons, of which the most important are finding food, avoiding attack by predators and migrating. In every case the primary motivation comes from the individual, for whom the advantages of joining a group outweigh those of remaining solitary.

open ocean or vultures in the African savanna. As a result, both seabirds and vultures are particularly adept at noticing concentrations of their fellow birds from a distance. For some birds, flocking can help concentrate their food source: pelicans work together to corral fish into shallow water, where they can all feed more easily.

Pelicans, like other waterbirds, often come together in flocks to find food

How does flocking help a bird find food?

Once one bird has discovered a good food source, another will soon spot it and join in. As other passing birds are attracted by the gathering, a flock may form. This behaviour is especially useful for species whose food tends to be scattered or concentrated in a small area, such as seabirds in the

How does flocking help a bird avoid being attacked?

The other main advantage of flocking – either while feeding, migrating or going to roost (see page 115) – is that it reduces an individual's chances of being attacked and killed by a predator. This comes down to basic mathematics: the more individuals there are for a predator to attack, the less chance a single bird has of falling victim. Another advantage is that a flock provides numerous eyes and ears to look and listen out for predators, and if one does appear, individuals can join together to mob it. Finally, rapid twisting and turning movements by flocks of birds such as waders or starlings is thought to confuse an attacking predator such as a Peregrine.

How does flocking help during migration?

Flocking is also advantageous for some, though not all, migrating birds: notably those that travel in flocks made up of family groups, such as geese and cranes. There are two main advantages. First, by flying in a V-formation they reduce wind

Snow Geese form vast flocks when on migration

Geese migrate in a V-formation

resistance, with birds taking turns in the lead position (rather like a team of racing cyclists). Second, travelling together reduces the chances of getting lost, especially for young birds on their first migratory journey. Having said that, many birds (especially songbirds) migrate either in loose flocks or on their own, and still manage to reach their destination successfully.

How do birds in a flock communicate?

Generally by call – as in the 'contact calls' made by tits when in a winter feeding flock, or the flight calls of waders. If a predator should approach, the first individual to spot it will often sound the alarm, warning other flock members of the danger and enabling the alert individual to escape in the resulting confusion.

Why don't birds in a flock collide with each other?

Fast-moving flocks operate on the principle that each individual bird is able to maintain its own 'personal space' – in effect a mini-territory that no other bird may enter. Flocking birds have an instinctive appreciation of the distance they must maintain from the bird in front and those on each side, even when moving at high speed, and the ability to take evasive action if these birds deviate from their course or slow down. This produces the 'twisting and turning' effect seen in vast flocks of ducks or waders, in which the flock appears to behave virtually as a single organism.

So how do the birds in a large flock all turn at the same time?

By a process similar to that of a 'Mexican wave', in which as one bird on the edge of the flock begins to turn, the bird next to it turns a split-second later, and so on. Birds have a lightning-fast reaction time and very quickly see an approaching movement, enabling them to anticipate it. As a result a wave can move through a flock in less than 15 milliseconds – to our eyes appearing almost instantaneous.

Are flocks always made up of a single species?

No – mixed flocks frequently occur. These usually comprise several closely related species, such as ducks, waders, thrushes or finches, but sometimes also include totally unrelated groups, such as winter flocks in woodland which may contain tits, nuthatches, creepers and even the occasional woodpecker. Again, the advantages of flocking, and the shared goal (e.g. finding food) outweigh any disadvantages such as competition.

Do all birds flock?

No. Some, such as most cuckoos, are almost always solitary; whether feeding or on migration. Most birds of prey and owls are usually found singly or in pairs, though there are exceptions, including the Red Kite and small falcons such as the Red-footed Falcon, both of which feed in loose flocks. This is because their prey consists, respectively, of carrion and flying insects, meaning that there is usually plenty to go round. Some raptors that are usually solitary will roost communally at night (harriers) or migrate in flocks (Honey-buzzards and several species of eagles).

Birds like these Starlings flock together to avoid danger from predators

What is 'mobbing'?

The attacking and harassment of a predator by a group of smaller birds, usually songbirds. This normally takes place when a predator such as a hawk or owl turns up unexpectedly in the birds' territory, or they discover it while out feeding. If the predator is flying, several birds may fly up to hassle it until it departs; if it is perched, they will fly down and attack it, calling noisily until it has had enough. Some mobbing birds are barely smaller than the object of their fury: crows, for instance, often mob Buzzards. Other kinds of predator may also provoke an attack, and birdwatchers searching for owls in the Tropics should bear in mind that small birds often mob snakes.

Crows often mob birds of prey, like this Brahminy Kite

Does mobbing work?

In most cases, effective mobbing will drive away a bird of prey, at least temporarily. However, if a bird gets too close it may fall victim to the very danger it was trying to prevent. Occasionally, another large bird, such as a passing heron, will be mobbed in error. (Waders mob herons fiercely, as herons will prey on their chicks.) On the other hand a predator that is a familiar part of the scenery, such as a breeding Peregrine, will often be left in peace, and only mobbed if it actually tries to attack and kill another bird.

What is 'roosting'?

Basically, taking a rest. Although this term is often used to describe large nocturnal gatherings of sleeping birds, it can also apply to just a single bird, can take place at any time of day or night, and is not necessarily a prelude to sleep.

Snowy Owls roost by day

Do birds always roost at night?

Certainly not. Nocturnal species such as owls roost by day, when they may sometimes be seen perched in a hole or crevice in a tree. Waders may roost by day or night: their feeding times are governed by the twice-daily rise and fall of the tide, rather than by the circadian rhythm which rules most other birds' lives.

Why do birds roost together?

Mainly to avoid attack by predators, just as flocking enables each individual to reduce its chances of being attacked. Roosting birds benefit from the rule of 'safety in numbers' by remaining collectively alert, and can respond quickly to alarm calls. Roosting close to other birds can also be a good way to keep warm.

Does roosting in flocks have any disadvantages?

Not many for the birds, though overcrowding may increase the risks of spreading disease. However, bird roosts can cause all sorts of problems for people: large gatherings of pigeons or starlings befoul building ledges with their droppings, which may carry dangerous diseases; and noise can be a problem in residential areas.

Flocks of Feral Pigeons can cause problems in some cities

6 · WHAT DO BIRDS EAT?
Feeding

EATING

What do birds eat?

A better question might be 'What *don't* birds eat?' There's hardly an edible item anywhere that doesn't pass down some avian gullet. Typical diets include insects (e.g. warblers and flycatchers), seeds (finches and buntings), meat (eagles, hawks and owls), fish (seabirds and herons), fruit (parrots and turacos), aquatic invertebrates and plants (ducks and geese), grain (gamebirds) and crustaceans (waders, gulls and seabirds). For the more sophisticated palate, there is also nectar (hummingbirds), lichens (Ptarmigan), tree sap (sapsuckers), rotting meat (vultures), wax (honeyguides) and even faeces (gulls, crows and vultures). If it is edible (and sometimes even if it isn't) some bird, somewhere, will eat it.

Eastern Bluebird with worm

of prey such as vultures will gorge themselves on a carcass and not feed again for several days afterwards.

How often do birds eat?

Feeding rates vary dramatically, from once every second or two (insect-eating songbirds) to just once a day (several species of grouse). Large birds

How much do birds eat?

This depends on the size of the bird. In general, the smaller and lighter a bird, the more it will eat in proportion to its body weight. Large birds such as

Herons feed mainly on fish, though they will also take amphibians and even small birds and mammals

raptors tend to eat around one-quarter of their body weight a day; while smaller birds such as tits and warblers may eat up to one-half – though in actual terms this may still only be a few grams. The amount birds eat changes from season to season too: small birds must eat more in cold weather in order to maintain their body heat, or they risk death. Long-distance migrants also feed much more intensely just before setting off, to build up fat reserves for the journey ahead.

Where do birds find their food?

The surface of our planet is one big buffet for birds. Some feed on the ground (e.g. Robins, thrushes and pigeons); others do so in trees (warblers, tits, parrots). Some catch their prey on the wing (swallows, swifts, falcons); others probe into the mud (waders). Some forage beneath tree bark (woodpeckers, creepers,

nuthatches); while others dive beneath the surface of a river, lake or ocean (dippers, ducks, gannets). If there is an ecological niche with food on offer, you can be sure that at least one species of bird has evolved to exploit it.

Puffins feed on sand eels, which they hold in their specially adapted bill

Hooded Vultures are mainly scavengers, feeding on carrion

How do birds find their food?

Each species has its own strategies, according to what it feeds on. Many simply forage around their chosen habitat, picking up morsels as they go. Others must hunt more actively: such as a swallow snapping up tiny airborne insects, or an eagle chasing and catching a hare. In some places, such as a rainforest tree canopy, food is everywhere. In others, such as the open ocean, it may be concentrated in one particular place, and take time and effort to find. So some birds, such as many woodland or farmland songbirds, usually feed within a relatively small area; while others, including seabirds and large raptors, may have to wander for hundreds of kilometres to find a meal. Many species, including New and Old World vultures, corvids and gulls, make a living out of scavenging dead or dying creatures; while in recent years they have also become adept at exploiting man-made food sources such as rubbish tips.

Which senses help birds find food?

Mainly eyesight – either for spotting food, or for watching the actions of their fellow birds and following them to it. However some species, such as nocturnal owls and nightjars, hunt mainly by hearing; while a few, including Turkey Vultures and storm-petrels, do so by smell. Turkey Vultures and their relatives have 'see-through' nostrils, which improve the flow of air through their nasal cavity and enable them to pinpoint the source of rotting food even more effectively. Finally, some birds that probe deep into the mud or earth, such as Snipes and Woodcocks, find their prey by touch, using sensitive nerve-endings at the tip of their bill.

Owls, such as this Barn owl, hunt mainly by using their hearing

Do birds chew their food?

No. Birds lack teeth (which would make them too heavy to fly), so they swallow their food whole. Some birds of prey tear their meat into chunks and strips to make this job easier, and finches will normally crush a seed in their bill, and discard the shell and husk before swallowing.

How do birds digest their food?

After swallowing, food passes down a bird's oesophagus, either into the 'crop' (a storage area in the throat) or directly into the first section of its stomach, where digestive juices begin to dissolve it. Afterwards it passes into the second section, known as the gizzard, whose powerful muscular walls grind hard substances – effectively doing the same job as a mammal's teeth. However, some food items such as fur and bones are impossible to digest, and are regurgitated as a 'pellet'. Owls are particularly well known for their pellets, which can often be found beneath trees at regular roosting sites.

Sugar rush

The birds that eat the most food (relative to their body weight) are undoubtedly the hummingbirds, which will consume about twice their body weight every day, mainly in the form of nectar (though also including some tiny insects).

Do all birds have a 'crop'?

No, some do not. Those that do include pigeons and doves, parrots, raptors and gamebirds. The crop, a pouch beneath their throat, allows them to store items for later digestion (when there is a sudden glut of food), or to bring food back to the nest and regurgitate it to feed their young. Storing food in this way also enables a bird to feed quickly, minimising the risk of attack.

Macaws obtaining minerals at a clay lick in the Peruvian jungle

Why do some birds lick salt or eat grit?

Many birds eat small items of grit in order to improve the grinding process of their gizzard. This habit is particularly common amongst seed-eating birds such as pigeons and doves, finches and buntings, gamebirds and Ostriches. Salt-licking, best known amongst certain species of parrot, may be a way of neutralising toxic substances in their food.

How do birds excrete?

Unlike us mammals, birds only have a single opening for excretion, the cloaca, which also has to handle reproduction and egg-laying. Before excretion, most of the liquid waste is reabsorbed into the body, so birds' waste products are generally fairly solid in substance – though the owners of bespattered cars may beg to differ.

Do birds co-operate to find food?

Some certainly do, either intentionally or unintentionally. Waterbirds such as pelicans and cormorants, and some ducks including Shovelers, will congregate in a suitable area and surround their prey, herding it into one concentrated mass where each individual can feed to its heart's content. Some woodland species may also co-operate, such as mixed flocks of tits, which call constantly to each other (though this may simply be to keep the flock together as a safety measure).

Young Little Owls beg their parents for food

How do young birds learn to find food?

Like young animals of all kinds, including human babies, by a combination of instinct, observation, and trial and error. The pecking instinct is innate, as can be seen in newly hatched chickens or gamebirds. A young bird learns as it grows, by observing the feeding actions of its parents and other birds in the group, and can soon distinguish between tasty and toxic. Some adult raptors, such as Peregrines, will actively help their offspring catch food by driving prey towards them.

A Great Tit feeds its youngster, even though the chick is now out of the nest

How have birds' bills adapted to their different diets?

Because birds deal with food using their bill, this appendage has evolved a wonderful variety of shapes and sizes to exploit the full spectrum of grub on offer. Taking just one group, waders, as an example: plovers have a short, stubby bill for picking up small items; Dunlins have a longer, thinner, down curved bill for probing beneath the surface; avocets have an upcurved bill for sweeping across the water to snap up tiny morsels; and Snipes have a long, thick bill for poking right down into the mud. Every wader species has evolved a different bill shape to deal with the particular food it eats (and in doing so, has provided birders with a good way to tell them all apart). Other amazing bills include those of hummingbirds, which are long and thin for probing flowers for nectar; flamingos, shaped to allow them to filter out tiny organisms using their tongue as a pump; and birds of prey, whose bills are hooked and powerful for tearing meat.

The Little Ringed Plover has a short bill to pick items off the surface of mud

The Dunlin has a short curved bill for probing for food

Avocets have an upcurved bill for skimming the surface of water

Do pelicans really keep food in their large bills?

"A wonderful bird is the pelican; his beak will hold more than his belly can…" This famous rhyme alludes to the extensible throat pouch, called a gular sac, which hangs below a pelican's bill and is used to 'net' fish rather than store them. Any fish caught will first be swallowed, then either digested or later regurgitated for the bird's young.

Flamingos have a very specialised bill for filter feeding

121

The North American Snail Kite has a very thin bill, which it uses to feed on snails

Do all birds have a specialised diet?

No. Some groups, such as many gulls and crows, will consume most edible items they come across; while others, such as hummingbirds (nectar), raptors (meat) and seabirds (fish) are best equipped for dealing with one particular kind. Some species are even more specialised: the Snail Kite of the Americas, as its name suggests, feeds almost exclusively on snails (apple snails, to be precise); though even this species will take other kinds of food when conditions force it to. Specialisation is a good way to avoid competition, and works well in times of plenty; but it does make the species vulnerable to any environmental changes that threaten its favoured food supply.

Do birds change their diet from season to season?

Many have to, since some foods are simply not available all year round. So in spring and summer, tits feed mainly on insects such as flies and caterpillars; but in autumn and winter, when these are scarce, they change their diet to include berries, windfall fruits and various items we offer them on bird-tables. Migrants also change their diet: stocking up on berries in autumn to provide energy for the long journey south. Many birds can also broaden their diet to take advantage of brief seasonal bonanzas: Tawny Eagles, for example, will drop everything in order to binge on termites, when these flying insects periodically emerge onto the African savannah in their millions.

Tawny Eagles sometimes forgo their usual diet to feed on termites

Why don't birds eat grass and leaves?

The Hoatzin of South America is able to feed on leaves, unlike other birds

Because unlike ruminant mammals such as cattle, deer and antelopes, birds are unable to digest them. Also the nutritional value of grass and leaves is very low, so a huge volume must be consumed in order to derive enough energy to be worthwhile. For flying birds, the weight of all this would seriously affect their ability to take to the air. However, one species, the Hoatzin of the Amazon rainforest of South America, does eat leaves as part of its diet. It is able to do so because unlike other vegetarian birds, in which food is broken up in the gizzard, the Hoatzin has a crop with thick, muscular walls and a horny lining. This enables it to crush the leaves before they are finally digested in the stomach. As a result the Hoatzin often appears 'front-heavy', especially after a large meal, and can only keep its balance when its crop is full by resting its head on its specially developed breast-bone.

Do birds eat other birds?

Yes. Apart from the obvious examples (day-flying birds of prey such as eagles, hawks and falcons), other bird-eating groups include owls, shrikes, and some storks and herons. Some birds of prey will feed on other raptors: the European Eagle Owl, for example, frequently takes Buzzards, and Long-eared and other owls. On occasion, a bird will be driven by starvation to feeding on other birds as part of its diet: in one particularly hard winter during the 1980s, a Water Rail was observed killing and eating Blue Tits in a Yorkshire garden.

What proportion of birds are predators?

That depends on how you define the word 'predator'. If we just count birds that feed mostly on warm-blooded creatures (i.e. other birds and mammals), it would only include diurnal raptors (such as eagles, hawks and falcons), owls and shrikes. These groups contain roughly 600 species in all (just under six per cent of the world's birds). However, if fish-eating birds such as Osprey, wildfowl, seabirds, herons and kingfishers were included, this total would rise to more than 1,000 species (ten per cent). And if catching and eating any animal prey counts as being a predator, then all insect-eating birds, such as flycatchers, warblers and Robins, should also count.

The South American Harpy Eagle is one of the biggest and strongest of all the world's raptors

Big game hunters

The dubious honour of the largest prey item killed by a bird goes (posthumously) to various monkeys and small antelopes, all of which regularly fall victim to some of the larger eagles. The Crowned Eagle of sub-Saharan Africa has even been known to kill a sub-adult Impala weighing 30kg (66lbs) – five times heavier than the bird itself. However, whilst these formidable predators are fully capable of killing animals larger than themselves, they cannot carry away anything that weighs more than they do. The largest wild animal known to have been killed and carried away by a bird is a 6.8kg (15lbs) male Red Howler Monkey, killed by a Harpy Eagle in Manu National Park, Peru, in 1990. This feat is all the more amazing given that the eagle itself would have weighed no more than 9kg (20lbs).

Are all raptors carnivorous?

No, though the vast majority are. Notable exceptions include the Honey-buzzard, which feeds mainly on the larvae of insects such as bees and wasps; and the Palm-nut Vulture of Africa that, as its name suggests, feeds mainly on the fruit of the oil palm. However, these species reflect their ancestral lineage by possessing all of the attributes we associate with birds of prey, such as a hooked bill, sharp talons and forward-facing eyes.

White-tailed Eagles specialise in catching fish

How do raptors catch their prey?

Mostly with their feet: grabbing or striking their target with powerful talons and sharp claws, and squeezing or tearing it to death. Even when catching flying insects such as dragonflies, falcons such as the Eurasian Hobby will grab them with their feet before passing them into their mouth with a single, smooth action. If feet do not finish the job, raptors will use their sharp bill to tear their victim's flesh. Falcons have a special notch in the upper mandible of their bill, called the tomial tooth, which is adapted for severing tiny vertebrae.

Do other birds use their feet to find food?

Yes, in various different ways. Herons and egrets 'foot-paddle', stirring up the mud at the bottom of the water in order to bring food items such as small invertebrates to the surface. Other birds, such as gulls, do a similar thing, by stamping their feet repeatedly on the surface of a grassy field – an action that appears to make worms and other creatures rise to the surface. The Secretary Bird, a bizarre bird of prey found on the African savannah, actually stamps on its prey – a very effective way of killing beetles, rodents, lizards, snakes and even ground-dwelling birds. One species of rail, the Purple Swamphen of southern Europe, Africa, Asia and Australia, grips aquatic vegetation in its foot and transfers it to its mouth as if using a 'hand'; while a similar action is performed by parrots when feeding on fruit.

The Secretary Bird uses its powerful feet to stamp on its prey

Do birds hide or store food?

Yes, in a habit known as 'caching'. In times of plenty, especially in early autumn when trees are fruiting, several species will hide away food in a 'cache' for the coming winter. This is common amongst members of the crow family, such as nutcrackers and jays, which bury acorns and other seeds under leaf litter. Their ability to find this food again after months have passed may show either intuition or intelligence, but certainly demonstrates an impressive memory. Another species that regularly stores the fruit of the oak tree is the aptly named Acorn Woodpecker. Found in the western USA and Mexico, this attractive woodpecker drills long rows of holes in the trunks and branches of trees, and crams an individual acorn into each one. Certain shrikes use a rather more grisly 'larder' to store their food. They impale their bird, amphibian and reptile prey on the spikes of thorn bushes, a behaviour that has earned them the nickname of 'butcher-birds'.

Do birds use other animals to help them find food?

Yes, in a process known as 'commensalism', which is defined as two species (plant or animal) living closely alongside each other without becoming interdependent. Thus European Robins will follow wild boars, in order to pick up worms and other

A Red-billed Oxpecker hitches a ride on the back of a Zebra

food items that they dig up while foraging. In recent times they have taken to following human gardeners for the same reason. Cattle Egrets and several species of gull and wader will follow large domestic mammals such as cows and horses, or wild mammals such as elephants and wildebeest, in order to take advantage of the insects they either disturb or attract. Some species, such as oxpeckers, ride almost permanently on the backs of large mammals, feeding on the insects attracted by the animals' body heat and sweat, and even using the animal's hair as nesting material! But perhaps the most extraordinary example is that of the antbirds, a South American family of passerines. Antbirds do not feed on ants, as might be expected from their name; instead, several species have learned to follow troops of giant army ants as they sweep across the ground, snapping up any insects they flush along the way.

Left: The incredible larder made by the Acorn Woodpecker of North America

Man-eater!

There is no shortage of lurid tales that describe immense eagles, especially Golden Eagles, snatching up small children. Such horrors can generally be dismissed as the stuff of myth and legend. However, there are a handful of records involving the Crowned Eagle, which – if genuine – make it possibly the only bird to prey on human beings. These records, which must represent extremely unusual circumstances, include the discovery of part of a child's skull in a Crowned Eagle's nest. Since this species never feeds on carrion, and often takes monkeys at least as large as a human baby, the unfortunate infant was almost certainly snatched while still alive.

Do any birds have symbiotic relationships with other creatures?

Symbiosis is defined as a relationship between two different species which brings mutual benefit to both, and is extremely rare in birds. The best-known example is the honeyguides, an African and Asian family related to the woodpeckers, whose family name, Indicatoridae, gives a clue to their unique skill. Two species of honeyguide are especially partial to honeycomb and beeswax, but are not always able to obtain this food without help, as the bees would attack them. So they have apparently learned to guide a honey-loving mammal, the Ratel or Honey-badger, to the bees' nest, using their call. The badger is immune to bee stings, and rips out the nest, allowing the bird to feast on wax and larvae in the remaining comb. Human observers have also learned to follow honeyguides, which have developed a symbiotic relationship with man in some places. (Local custom encourages people to always reward the bird by leaving it a portion – otherwise next time it might take its revenge by leading them to a venomous snake.) It might also be argued that garden birds coming to feeding stations have developed a similar relationship: we give them food; they give us pleasure.

Do some birds steal from others?

Yes. This process is known as 'kleptoparasitism' – or as many birders call it, 'piracy'. Many birds will opportunistically snatch food from another – just watch the thieving at a gull colony. But two groups of seabirds have turned a life of crime into a fine art. Skuas (jaegers) and frigatebirds are the pirates of the air, pursuing and harassing other birds until they drop their food, then catching it acrobatically in mid-air. Skuas' victims include terns and gulls, while frigatebirds often pursue elegant tropicbirds in a real 'beauty and the beast' contest. Large raptors such as the Bald Eagle and White-tailed Eagle also sometimes try their hand at piracy, their victim of choice often being the unfortunate Osprey.

Vultures are classic scavengers, feeding on carrion

What is a 'scavenger'?

Any species that feeds primarily on dead meat, such as Marabou Storks, and both New and Old World vultures. Some of these birds have bare, unfeathered heads to prevent the clogging of feathers with blood and gore. Other species, such as Magpies and crows, have a more varied diet, but won't turn up their noses at a carcass when opportunity allows. Most gruesomely of all, Turnstones have been recorded feeding on a human corpse washed up on the tideline.

Skuas often steal food from other birds

How do vultures avoid food poisoning?

By having very strong stomachs! Vultures have a highly specialised digestive system, with powerful acids to neutralise even the most putrescent rotting meat that would quickly kill a human being.

The Sharp-beaked Ground-finch of the Galapagos uses its bill to feed on the blood of nesting seabirds

Going large

The biggest food item eaten by any land bird must surely be a dead African Elephant, which may draw hundreds of vultures of various species to binge on its bloated carcass. But this is a mere snack beside the beached whales on which scavenging seabirds such as the Southern Giant Petrel and Antarctic Skua gorge themselves whenever they get the opportunity.

Are any birds true parasites?

True parasitism – defined as the one-way exploitation of one living organism (the host) by another (the parasite) – is very rare amongst birds. Examples include the Sharp-beaked Ground-finch (or 'Vampire Finch') of the Galapagos, which regularly takes blood from the base of the feathers of Masked and Red-footed Boobies; and Kelp Gulls, which have been known to peck at the sores on the backs of Southern Right Whales to eat the flesh.

A Kelp Gull stealing a penguin egg

DRINKING

How do birds drink?

Most birds have to tip their heads back to allow water down their throat, as they lack the ability to swallow. However, a few groups of birds, including pigeons and doves, the African mousebirds, and some species of finch, are able to drink without lifting their heads, using their tongue to create a sucking action. Other species such as swallows and martins fly low over water, dipping their bill just below the surface to grab a billful of liquid as they pass. In the Antarctic, penguins drink by eating snow.

Swallows are able to drink on the wing

Where do birds find water to drink?

In all kinds of places: some obvious, such as lakes, puddles and ponds; others less so, such as dew, or rainwater on leaves or in the base of plants. Some species drink falling rain directly. But many, especially insect-eaters, get most of the liquid they need from their food.

How often do birds need to drink?

This varies considerably, depending on their diet. Insect-eating birds such as warblers may only have to drink once a day, whereas birds that feed on seeds and other dry foods will have to do so at least two or three times. Birds that obtain all their liquid requirements from their food do not have to drink at all.

How do young birds in the nest get water?

Usually their parents carry water in their bill and then pass it drop by drop into the mouths of the nestlings. However, desert-dwelling sandgrouse have evolved a unique way of carrying back

Sandgrouse carry water back to their young by soaking it up in their specially adapted breast feathers

more water in very dry conditions: the male soaks his breast in water, and then flies back to the nest where the thirsty young are able to get moisture from his uniquely-adapted, sponge-like feathers.

How do seabirds drink salt water without coming to harm?

Although most birds cannot drink salt water, ocean-going seabirds such as albatrosses and shearwaters and some others have a specially adapted salt-gland in their nasal region, which enables them to excrete the high levels of salt found in seawater through their nostrils. As a result they can stay out in the open ocean for weeks, months, or even years, on end.

Seabirds such as the Manx Shearwater have a special bill which enables them to expel salt from seawater

GARDEN BIRDS

Is it good to feed garden birds?

In general, yes. Feeding prolongs the lives of individual birds and helps maintain numbers. Without supplementary food, many birds would die, especially during hard winter weather. With so much damage and destruction in other habitats such as farmland, gardens have now become important refuges for many species – especially songbirds. Feeding birds also brings tangible benefits for us: it is a hobby enjoyed by millions of people, especially in Britain and North America, and leads many onto a more active interest such as birding or conservation.

Can feeding birds do any harm?

Possibly, yes. Concentrating birds in one area can help spread diseases. Also, it may tip the balance in favour of predators such as raptors or cats, by concentrating birds in one area, where the attacker can claim a 'free lunch'. A more subtle argument against feeding is that by artificially helping birds we are acting against nature, and supporting individuals which otherwise would have died. As a result, it is argued, the population as a whole becomes less 'healthy'. However, it might also be argued that we have destroyed so much natural habitat that we have a moral duty to lend the birds a helping hand. Overall, the benefits of feeding appear to outweigh the drawbacks.

Blue and Great Tits at a bird feeder

What are the best types of food to give birds?

Birds feed to gain energy, so the best foods are those that deliver the purest form of energy in the quickest time. For this reason, sunflower seeds are better than peanuts; while shelled 'sunflower hearts' are best of all, as the bird does not need to spend valuable time removing the kernel from the husk. Fat, in the form of fat balls or suet, is another high-energy food, especially valuable in winter. Live food such as mealworms can also be a vital supplement during the breeding season, when nestlings and parent birds need all the help they can get.

Should I feed birds all year round?

Yes, definitely. Despite the long-held belief that we should only feed birds during winter, scientists have now proven that feeding brings benefits to birds all year round. The most important times of year to feed are spring (when parent birds must lay eggs and feed young), and during any prolonged period of harsh winter weather.

Bird baths and nestboxes don't need to be elaborate – any raised dish or homemade box will do

How else can I attract birds to my garden?

There are four other areas to focus on if you wish to attract more birds – and a greater variety of species – to your garden. First, provide water as well as food; it is essential for drinking and bathing. A birdbath is great, and a pond even better. Second, provide a place to nest: either in the form of suitable shrubs and bushes; or the quick and easy way, by putting up nestboxes. Third, guard against diseases, by keeping your feeding station clean and tidy; and against visiting cats, by making them unwelcome! Finally, by growing a selection of native plants – including flowering annuals, bushes and shrubs, and trees – you will provide natural food in the form of nectar, berries, fruit and insects for the birds to enjoy.

Grey Squirrels are often unwelcome visitors to bird feeders

How can I keep squirrels at bay?

Some people adore the antics of squirrels; others detest them. Before you begin an all-out war you might consider taking a more relaxed attitude; but if you really want to stop squirrels eating you out of house and home there are several fairly effective products on the market that have been specially designed to allow small birds to feed while keeping squirrels out. And if you do decide to invest in a 'squirrel-proof' or 'squirrel-resistant' bird feeder or table make sure you site it well away from any trees or fences, or anything else in your garden that a squirrel might jump from to try and get on top.

How else do birds take advantage of humans for food?

Apart from the vast amount we spend on deliberately providing food for birds, our modern lifestyles and wasteful habits provide unintentional rich pickings. Birds have always followed human hunters and fishermen in order to scavenge carcasses or pick up any waste scraps; but in the modern world they have a vast range of other opportunities to take advantage of us. No landfill site (a euphemism for rubbish tip) would be complete without its hordes of gulls, crows and – in tropical regions – kites and vultures, scavenging for scraps of waste food. Grain stores are also popular amongst seed-eating birds, though in recent years less wasteful storage methods have reduced this supply. Seabirds will follow fishing boats to feed on scraps and waste cast overboard, while passenger and cargo ships will churn up food items from beneath the surface for birds to pick up. Even our modern transport system provides a steady supply of roadkill victims for enterprising species such as crows and kites.

Gulls waiting for scraps from a fisherman

7 • WHY DO BIRDS SING?
Communication

Why do birds sing?

For us, birdsong may be a thing of beauty; for birds, it is all about love and war. War, because one reason birds sing is to defend their territory against rivals. Love, because the other reason is to attract a passing female, and persuade her to mate.

Why do they use sound, instead of showing off their plumage?

Lots of birds do indeed use visual signals. For example, colonial nesters such as seabirds perform elaborate displays to warn off rivals and attract a mate. But species living in woods or forests, or in the middle of a dense reedbed, are often hidden from view, so birdsong evolved as the best way to communicate over long distances. For songbirds, sound also enables them to communicate in the period before sunrise, in the celebrated 'dawn chorus'. Many species, from Capercaillie to birds-of-paradise, use both sight and sound to attract a female.

Displaying male Capercaillie

Can all birds sing?

No. Apart from the 6,000 or so species of songbird, most birds cannot sing, though this does depend on how 'singing' is defined. Many non-passerines have special vocalisations in the breeding season that are quite different to their calls. For example some waders such as the Green Sandpiper sit on an exposed perch and deliver what sounds very like a song.

What is the difference between a song and a call?

Songs are more complex series of notes or phrases, uttered mostly by male birds, usually in spring, and used to repel rival males or attract a mate. Calls, on the other hand, are short, simple utterances, produced by both males and females throughout the year, for a variety of reasons. In addition songs are largely confined to the group of birds appropriately known as 'songbirds'. Generally, songs are also more attractive to the human ear than calls: thus the Blackbird's deep, fluty and tuneful song is justly celebrated; while its repetitive, clanging alarm call is not. But there are many exceptions: some songs, like that of the Chiffchaff, are as simple and repetitive as the calls of other species.

A Chiffchaff singing in spring

Name that tune

The widest repertoire of any bird is that of the Brown Thrasher of North America. This member of the mockingbird family has as many as 2,000 different songs, about 10 times the number written by the Beatles. This is far and away the largest 'songbook' of any bird: distant challengers include the Nightingale (between 100 and 300 different songs), Song Thrush (between 140 and 220), and Northern Mockingbird (up to 150).

The Brown Thrasher has an extensive repertoire of different songs

Are any birds completely silent?

Although most birds have some kind of song or call, a few groups, such as New World vultures and storks, lack the vocal mechanism to make complex sounds, so apart from the occasional hiss, grunt or croak are usually silent. Incidentally, the name 'Mute Swan' arose because this species does not call in flight, unlike Bewick's and Whooper Swans. Mute Swans do hiss though, especially if you get too close.

Why don't females usually sing?

Because they don't usually have to defend a territory, or take the lead in courtship – though in fact the female's role in choosing a mate is much more active than we once thought. There is also a physiological explanation as to why females seldom sing: they don't usually have such high levels of testosterone, the hormone that triggers male song at the start of the breeding season.

Is it only males that sing?

By and large, yes, though there are exceptions, such as the European Robin. In spring the male guards his territory just like any other bird, but in autumn and winter both males and females have their own separate feeding territories, which they defend against rivals of both sexes by singing.

A male Blackbird in song

What message does a singing bird send to his rivals?

A male singing in his territory is sending a clear message: "this is my patch, so clear off!" The louder and more frequent the song, the more likely a rival will get the message. But in species where territories are fairly close together, such as the Blackbird and Song Thrush, the male will leave distinct gaps between phrases so that he can detect any answering male in the vicinity.

Robins are one of the few species where both males and females sing

Is song always an effective deterrent?

No. A rival male who does not yet have a territory of his own may decide that 'fight is better than flight'. Instead of fleeing he will enter the territory and begin singing himself: a clear signal to the occupier that he poses a threat. The two rivals will then take part in a 'knockout competition', in which the winner will usually be the bird with the 'best' song. Just like an avian talent show, but in this case the loser votes himself off.

A Robin attacking a Dunnock which has ventured into its territory

One hit wonders

The smallest repertoire of any bird is shared by several species, including the Ovenbird and White-crowned Sparrow of North America, and the European Redwing. These species have but a single song, with no variants – though this is no less effective at doing its job.

Does the rivalry ever go further?

In a few cases, rival males will resort to physical violence, with one bird attacking the other until either the occupier or the intruder retreats. European Robins are particularly aggressive, with males occasionally fighting to the death.

How do males avoid a fight?

The occupying bird usually wants to avoid physical aggression, so he adopts a range of tactics designed to deter any possible rival. This generally involves singing as loudly, and for as long as possible, sending the signal that he is not to be messed with. He will also patrol the borders of his territory, singing at several points around the boundary, often on exposed song posts where he can be easily seen. These strategies create the impression that the territory is larger than it actually is so that other males are put off coming near.

Do males ever change their songs?

Yes, sometimes. Males in the African family of robin-chats will change their song in order to outshine rival males, so that if an intruder sings the same song as the territory-holder, he will switch to a new one. If the rival imitates that, he will switch again, and so on.

How does a male's song attract the female in the first place?

Like humans, birds attract a mate in a number of different ways, of which song is only one. A passing female is more likely to stop and check out her potential mate if the song is frequent, persistent, varied and in some cases complex, as this is an 'honest' signal of fitness.

A young Barn Swallow calls to its parent for food

Is a male with a 'better' song more likely to attract a mate?

Yes, but this depends on what is meant by 'better'. Some birds, such as the Chiffchaff, have to our ears a fairly monotonous song, which they repeat with little or no variation. Others, such as the Willow Warbler, have a more tuneful (and to our ears, attractive) song. At the greatest extreme, the Nightingale utters an extraordinary variety of notes and phrases, often for hours on end. But human perceptions of tunefulness or originality are irrelevant. What is important within any single species is the quality of each individual's song, which may be measured in volume, frequency or complexity. This is what indicates the bird's breeding potential, and is the key factor in influencing a female's choice of mate.

Once he has won a mate the male Sedge Warbler stops singing

Once a male has attracted the female, does he stop singing?

Sometimes, yes – especially if the species is monogamous (i.e. males and females are faithful to each other throughout the breeding season). So, for example, a Sedge Warbler sings constantly when trying to attract a mate, but once he has won her over he stops and devotes his energies to building a nest and raising a family. Singing, after all, takes time and effort. Others, especially polygynous species such as the Corn Bunting, where males mate with several females, carry on singing in order to defend their territory against any intruding males.

Why do birds sing mainly at dawn?

There are several reasons why the peak of birdsong occurs so early in the day, during what is celebrated as the 'dawn chorus'. First, it is still dark, so rather than waste energy on trying to find food that is difficult to see, a bird is better off marking its territory.

Second, females are often at their most fertile at dawn, so the male must make a special effort to guard against a rival mating with 'his' female. Finally, it is much easier to hear birdsong at dawn: not only because in our towns and cities there is less interference from traffic noise, but also because weather conditions are generally better at this time of day, with less wind and air turbulence. There is also a phenomenon known as 'temperature inversion', in which a layer of cold air is trapped close to the ground by a warm layer above. Sounds reflect off the boundary between these layers, allowing birdsong to carry further.

Rise and shine!

Judging the best dawn chorus is obviously subjective, but it can be argued that temperate latitudes across Europe and North America, with specific seasons and a slow, gradual sunrise, produce the best conditions for early morning song. Many would claim that if you want to hear the most intense and varied dawn chorus in the world, it is hard to beat an English woodland on a fine, still morning in May.

Do they sing at other times of day?

Yes, some males go at it all hours – especially early in the breeding season when they have not yet attracted a mate and their territory may still be at risk from rivals. Generally, though, song activity declines during the middle of the day (especially during hot weather when it requires more energy), and reaches a peak again in the early evening although the 'dusk chorus' is rarely as loud, nor quite so intense, as its dawn counterpart. At dusk a similar temperature inversion occurs as at dawn, allowing the sound to carry further.

Why do some birds sing at night?

Some of the same reasons apply: it is quieter at night-time, the air is often more still, and most songbirds are unable to feed. Those species that do sing at night face less competition from other species: unlike dawn, when

Many birds sing at night, especially when there is a full moon

the whole orchestra is performing at full volume, nocturnal singers are more likely to find themselves doing a solo, without so many rivals. Modern technology may also have an influence: European Robins often sing at night in towns and cities, where they are illuminated by street lights, which may stimulate them to believe that dawn is approaching. Without doubt the most famous nocturnal songster is the Nightingale, which can be heard in much of Europe singing at full volume throughout the night from late April to June. This habit may have possibly evolved because males arrive back from their African winter-quarters a few days earlier than their mates, so by singing after dark they are more likely to attract returning females during their night-time migration.

Did a Nightingale really sing in Berkeley Square?

Almost certainly not, since the Nightingale is strictly a woodland species and steers well clear of Central London. The singer in question is much more likely to have been a Blackbird, Song Thrush or (by far the most likely) Robin. Perhaps inspired by the famous song, however, Margaret Thatcher is reported one February to have insisted to a senior official that she had heard a Nightingale singing outside number 10 Downing Street. The aide, a keen birdwatcher, bravely contradicted the Prime Minister, explaining that, since the Nightingale is both a woodland bird and a late spring migrant, she must have been mistaken. The Prime Minister persisted, and as the intrepid official was about to risk his career by

The Nightingale is probably the world's most famous songster

contradicting her for a third time, her Private Secretary discreetly intervened: "If the Prime Minister says she heard a Nightingale," he hissed, "then she heard a Nightingale!"

Loud and proud

The award for the loudest song of any bird is shared by the four species of bellbird, from the New World cotinga family. Their calls peak at up to 100 decibels, and can be heard from up to one kilometre away. The Superb Lyrebird of Australia also has an extremely loud call. However, neither can compare with the world's loudest insect – the male cicada, which at 150 decibels is louder than a jet plane passing directly overhead.

Holding a tune

The record for most songs in a single day goes to the Red-eyed Vireo of North America. One male was apparently reported as uttering its brief song-phrase a staggering 22,197 times in ten hours – on average once every 1.62 seconds. It is not known whether any females were suitably impressed.

Do birds sing all year round?

Generally, no. In temperate regions such as Europe and North America most birds breed during a period from February–March to June–July, roughly coinciding with the northern spring. Hence the gradual build-up of song from late winter onwards, stimulated by the steady increase in day-length. This reaches a peak in May when both resident and migrant birds are in full swing. Nevertheless, birdsong can be heard as early as November and December, especially in mild winters; or as late as August and September, towards the tail end of the breeding season. Some species, such as the European Robin, sing throughout the autumn and winter as well. In the southern hemisphere, roughly the reverse is true, while in the tropics, the conventional four seasons do not apply, so breeding activity is either spread throughout the year, or timed to coincide with periods of rainfall and the availability of food.

Do birds sing the same song all the time?

No. In general the song used to defend a territory is more direct and less complicated than that used to attract a mate. This is because the purpose of the territorial song is simple: to advertise that a particular area is already occupied; while the mating song needs a few more flourishes in order to persuade the female that this particular male is a 'quality' bird – i.e. one worth pairing up with.

How do birds sing?

They produce sounds through an organ in the throat called the syrinx, which is unique to birds. Air is exhaled from the lungs through the syrinx, where it passes across membranes, which vibrate and produce sound. Using various pairs of muscles attached to the syrinx, the bird can then vary

its song in four different ways: pitch, tone, rhythm and volume. By doing so, birds as a whole are able to produce an incredible variety of sounds, ranging from a simple monosyllabic call to an avian aria of amazing complexity.

The Gray Catbird has a remarkably complex song

How do birds manage to sing continuously without appearing to take a breath?

Like human singers and wind musicians, birds have evolved an extraordinary degree of control over their breathing, possibly by taking frequent 'mini-breaths' rather than large intakes of air, which would prevent or interrupt their song. A singing Canary may take up to 30 mini-breaths per second.

Can birds make different sounds at the same time?

As well as being able to vary pitch, tone, rhythm and volume, birds have another trick up their sleeve – or down their throat. They can produce sound simultaneously from both sides of their syrinx, which enables them effectively to combine two different sound channels into a single song. So the Gray Catbird of North America is able to emit six separate sounds in half a second, three from one side of the syrinx and three from the other, which combine to create the illusion of continuous singing.

How do birds create such variety in their songs?

Most songbirds use a 'kit of parts' to construct their songs. This comprises a basic template onto which they add their own variations, copying notes and phrases from their neighbours, and in some cases, mimicking other sounds they hear. Some species, such as the Chaffinch, sing a relatively limited number of phrases, usually uttered in the same order. The basic rhythm is fairly easy to recognise. One of the most accomplished songsters, the Nightingale, combines as many as 300 different song types in virtually random sequence – much like a jazz musician improvising on a theme.

How do birds optimise the effect of their song?

Getting your song heard by the largest number of potential rivals and mates is an important factor in breeding success. So some birds perch on top of a tree or bush, or in urban areas on the roof of a building. Where this is not possible, as in areas of open grassland, birds have found several different ways to broadcast their message more effectively. Some, like the Skylark, deliver their song in a long, continuous flight; others, such as the Meadow Pipit, launch themselves into the air with every short burst.

Do birds in different habitats have different types of song?

Yes, because songs have evolved to best exploit the acoustic properties of the bird's normal surroundings. So forest-dwelling birds tend to have deep, varied, tuneful songs, because high frequencies are absorbed by foliage; while in open areas such as reedbeds or grasslands, more monotonous, buzzy songs tend to work better. The Grasshopper Warbler, with its continuous high-pitched buzzing (similar to that of a fishing reel), is a good example of a song perfectly suited to its environment of low, scrubby vegetation.

The Grasshopper Warbler sounds rather like an insect – hence its name

Boom boom!

The farthest carrying sound belongs to the Bittern, whose booming call can be heard up to eight kilometres (five miles) away, due to its very low frequency. The Kakapo, a nocturnal, flightless parrot native to New Zealand, tramples a hollow in the ground and uses this to amplify its call. Thanks to this 'booming bowl' the call can be heard at least six kilometres (3.75 miles) away.

How do birds learn to sing?

A male Chaffinch (left) greets his mate

Scientists have long debated whether birds inherit their ability to sing, or whether they learn their repertoire by imitation and practice. As with human language, the answer is that they do both. Like us, birds are born with a basic ability to vocalise, but need to be exposed to full-blown song in order to learn it properly. So captive birds kept isolated from their fellow songsters develop a limited version of their 'proper' song; while those exposed to the song of a different species may adapt elements of it into their own song pattern.

Do different individuals of the same species have different dialects?

Yes. Chaffinches from different parts of Britain have quite distinctive differences in their song; and though these may be hard for the human ear to detect, they have the effect of only attracting females from the same local area. There may be a good reason for this: a local female is more likely to be adapted to breed in local conditions than is a passing stranger. By singing in the local dialect, therefore, a male maximises his chances of attracting a suitable mate. Eventually a particular dialect may differ so much from the original that its singers become reproductively isolated from neighbouring populations; and after time may even evolve into a separate species. This is what seems to have happened with the Iberian Chiffchaff of Spain and Portugal, which has recently been separated from its commoner cousin.

Can birds recognise other individuals of their own species?

Yes. It is thought that a singing male can recognise the songs of neighbouring males, and as long as they stay put on their own territory he will ignore them. However, if a stranger begins singing nearby a male will go into attack mode, seeking out the intruder and chasing him away.

Can birds understand the songs and calls of other species?

One way in which a species can co-exist with a similar species living nearby is by sounding different from it. Hence the 18th century naturalist Gilbert White's discovery that three similar species of 'leaf-warbler', Willow Warbler, Wood Warbler and Chiffchaff, have completely different songs and are thus able to tell each other apart. Nevertheless, some species appear to influence others with similar songs: thus Blackcaps and Garden Warblers living in the same wood may begin to adopt phrases and characteristics of each other's song; as may Reed and Sedge Warblers living in the same reedbed. Alarm calls are also readily understood by many different species: when one bird sounds a warning against an approaching predator, all small birds in the vicinity will take heed and dash to safety.

Reed Warblers (left) and Sedge Warblers (right) have similar songs, but the Sedge Warbler's is less rhythmic and more excitable

Do different calls have different purposes?

Calls fall into several categories. These include 'contact calls', for keeping in touch with others in a feeding flock, and 'alarm calls', for giving warning of an approaching predator.

Do any birds sing duets?

Some species do. Duetting is very rare amongst songbirds in temperate regions, but is found in a variety of tropical birds. This is because in the tropics many birds maintain a year-round territory, which may be packed with breeding pairs of a wide range of species. In the resulting cacophony, a singing male cannot be sure that his mate can hear him; so duetting

evolved as a way of keeping in contact. Singing duets is particularly common amongst African shrikes such as the Tropical Boubou, in which the male and female sing different 'parts'. Their timing is so perfect that to our ear the result often sounds like a single bird.

Do birds mimic other species?

If imitation is the sincerest form of flattery, then certain families of bird are the true sycophants of the animal kingdom. These include the starlings and mynas of Europe and Asia, the mockingbirds of the New World, the robin-chats of Africa, the lyrebirds and bowerbirds of Australasia, and various species of parrot from around the world. The ability to mimic appears to have evolved simply because it appears to make the singer more attractive to potential mates. Poor mimics simply don't make the grade, whereas good mimics do, and are able to pass on their skills to the next generation.

The Tropical Boubou sings duets with its partner

Can birds mimic artificial sounds?

Yes, they can do this too – despite the fact that most of these sounds are relatively new. For example European Starlings are accomplished mimics of recent technological innovations such as mobile phones and car alarms – much to the annoyance of people who would prefer a quiet life! The lyrebirds and bowerbirds of Australasia take this to even greater extremes, producing convincing impressions of chainsaws and other machinery.

Cheep imitation

Picking out the world's best mimic is virtually impossible: the Marsh Warbler has been found to mimic the calls and songs of more than 200 European and African species (the average individual can manage 70–80 different species); the African Grey Parrot can mimic several hundred words of human speech; while the Superb Lyrebird of Australia can imitate anything from a camera shutter to a chainsaw.

How can birds reproduce human speech?

By the same process as they mimic other bird songs and natural sounds. Oddly the best known imitators of human speech, the parrots, have rarely been observed mimicking sound in the wild; yet in captivity one African Grey Parrot, named Alex, had a vocabulary of more than 800 words. Parrots have a thicker, more muscular tongue than most birds, adapted for manipulating food in the bill, which may also make them more articulate.

Do they understand what they are saying?

General opinion would claim that they don't – hence the phrase 'parrot-fashion', meaning mindless imitation. Yet Alex the African Grey was taught not only to learn words, but could also combine them into simple commands, and even learned to say 'no' when refusing something he didn't want. Linguists are now rethinking the nature of avian mimicry as a result.

Do birds sing in the rain?

That depends on how hard it's raining. Light rain will inhibit some birds, especially those that live in open habitats, or that deliver their song from an exposed perch or in flight, but it won't usually bother those shielded by thicker vegetation. Heavier rain, especially accompanied by strong winds, usually prevents most birds singing: there's not much point, since a downpour would

drown them out – in every sense. But there are a few exceptions, notably the Mistle Thrush, known as the 'stormcock' for its habit of singing just before, or even during, foul weather.

How do birds that can't sing communicate?

Many birds outside the 'songbird' category make vocal sounds that fulfil exactly the same function as a song but are less complex. A male Bittern 'booms' from deep within a reedbed to advertise its presence, while most doves and pigeons repeat a very simple phrase that is effectively a song.

What other sounds do birds use to communicate with?

While songbirds take lead vocals, some other species are happier on percussion! The best known are the woodpeckers, which drum their bills on hollow wood in order to defend a territory and win a mate. Other non-vocal alternatives to song include the bill-clattering used by White Storks in their courtship display, and the Snipe's 'drumming': a rhythmic whirring sound made by spreading its tail and flying very fast through the air, which vibrates through the feathers. In the tropics, several species can produce sounds by trapping air beneath their wings and releasing it in an explosive 'click'. These include the White-bearded Manakin of South America, and various larks and cisticolas in Africa.

Woodpeckers such as this Great Spotted drum, rather than singing, to defend a territory and attract a mate

Female Mallard

Do any birds use 'echolocation', like bats?

Only a select few, all of which live in caves. The South American Oilbird (a relative of the nightjars) and several species of cave swiftlet from Asia have learned to bounce their calls off solid surfaces in order to locate flying prey or avoid bumping into cave walls. However, they are not nearly as accomplished at this as bats. Echolocation in birds is at normal frequencies, audible to the human ear, rather than at the higher frequencies that bats use.

How many birds are named after their sound?

Far too many to count! Some species have names that derive directly from the sound they make, including such well-known examples as Cuckoo, Chiffchaff, Kittiwake and Chough (originally pronounced 'chow' rather than 'chuff'). Others have names that describe their sounds, such as trumpeter finch, screech owl or whistling duck. The best example of the latter must be the African cisticolas – a family of very similar-plumaged birds whose common names reflect the importance of song in telling them apart: hence Rattling, Singing, Croaking, Wailing, Churring, Tinkling, Chirping, Trilling, Bubbling and Chattering Cisticolas, to name but a few. Some people even assert that the Rock-loving Cisticola is named not for its chosen habitat, but after its musical tastes.

Do all ducks quack?

Certainly not. The Mallard, which does quack, is the world's commonest and best-known species, so its call has come to represent all its kind. But the duck family as a whole has a much broader repertoire of calls. These include a high-pitched whistle (Eurasian Wigeon), a chorus of wailing cries (Long-tailed Duck), and a call that sounds like an elderly lady in a state of outraged curiosity (Common Eider).

153

8 • HOW DO BIRDS REPRODUCE?
Breeding

PAIRING UP
Do all birds breed in spring?

Springtime is certainly the season of love for birds in temperate regions – 'spring' being loosely defined as the period between the Spring Equinox and the Summer Solstice. In the northern hemisphere, therefore, this means a peak of breeding activity between March and June; and in the southern hemisphere between September and December. However, there are plenty of exceptions to this rule, even in temperate regions, where species such as Blackbird and Song Thrush may start breeding well before the 'official' start of spring – even in November or December, if the weather is particularly mild. And in tropical regions, breeding is generally timed to coincide with seasonal rainfall, so that there will be enough food for the young.

Sparrowhawk at nest with chicks

Why do birds breed at one particular time of year?

Birds time their breeding to coincide with the peak availability of food, which is why in Europe and North America most do so between March and June. The abundance of food at this time is especially important for species that feed on a single main type

Common Crossbills feed on pine seeds

of prey, such as Great Tits (caterpillars) and Barn Swallows (flying insects). This also suits the breeding cycle of birds of prey such as Sparrowhawks, who make sure they have chicks in the nest when there are plenty of young songbirds for them to eat.

Why do some birds breed earlier or later than others?

Breeding outside the usual season is related to the supply of a particular food. Common Crossbills start nesting in January or February, so that their chicks hatch during the peak fruiting period of the spruce cones on which they feed. In contrast, Eleonora's Falcons do not breed until late summer, with chicks in the nest in September and October. At this time, large numbers of migrating songbirds are passing through the Mediterranean region: ideal food for the young falcons.

How do birds know when to breed?

Birds cannot possibly 'know' when the food supply will reach its peak. Their decision to begin the whole process of breeding is usually governed by changes in day-length. These stimulate chemicals in their brain and trigger the cycle of courtship, finding a territory and so on. However, birds can also be fooled by periods of very mild weather in late autumn or winter, which may stimulate some to begin breeding at the 'wrong' time of year. Sadly this can have disastrous consequences, as a hard frost or heavy snowfall may wipe out the food supply just when the chicks need feeding, so that the young birds starve to death.

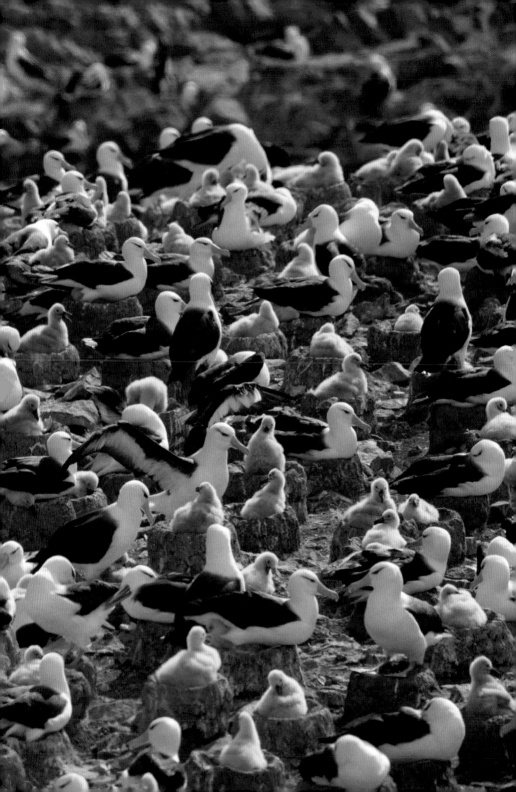

Opposite: Like most seabirds, Black-browed
Albatrosses nest in large, noisy colonies

Do all birds breed every year?

No. Although the majority of species,
especially those in temperate regions,
do breed annually, there are many
others that do not. Larger seabirds such
as albatrosses have a two-yearly cycle,
as it takes so long to raise their single
huge chick that they could not possibly
breed every year. Other seabirds such
as the Sooty Tern may breed more
frequently, for example on a nine-month
cycle. And some tropical seabirds and
penguins simply breed whenever it suits
them, so in any one colony there will be
young of all ages at the same time.

The male Peafowl displays to his mate

How do birds choose a mate?

The rituals of bird courtship are almost
infinitely varied and complex, but the
basic principle is always the same –
namely that one bird (almost always
the male) tries to dazzle his way into
his partner's affection by showing off
his wares. Tactics can include visual
displays, as used by colourful birds
such as the peacock (male Indian
Peafowl); vocal displays, as used by
songbirds such as the Nightingale;
and more complex 'behavioural
displays', such as the dancing of
cranes or food-passing between
raptors. Sometimes it involves a
combination of all three.

What is a 'territory'?

Simply an area defended by one bird
(almost always the male) or a pair in order
to carry out the processes of breeding
(courtship, copulation, nest-building and
raising young) without interference from
rivals of the same species.

Young lovers / late bloomers

The earliest age for a bird to begin breeding is between five and six weeks, in the
case of several members of the quail family, which can breed in the same spring
that they were born. The latest age for a bird to begin breeding is normally six to
10 years, for the larger albatrosses, though some Sooty Terns may not breed until
they are 10 years old.

Do all birds have territories?

The male Black Grouse in full display

That depends on how the concept of 'territory' is defined. Colonial species such as herons or seabirds may nest almost touching each other, but will still peck angrily at any intruder who dares to encroach on their 'personal space'. Species that use leks (see page 165) do not have territories as such, but will still jostle for the best position in the lekking area in order to attract the most females.

Do birds defend territories outside the breeding season?

Yes, because territory is not always connected with breeding. Birds will also protect a food supply during the autumn and winter months, when food may be scarce. Mistle Thrushes defend berry bushes against all-comers, while both male and female

European Robins maintain a non-breeding territory, using song to repel invaders. Hummingbirds may also defend a particular group of flowers against others, in order to keep a nectar supply for themselves.

Why do birds need territories in the first place?

For many birds, especially songbirds, a territory is a vital asset in the race to reproduce. By hanging on to his own territory during the breeding season, the male can maximise his chances of attracting a mate and being able to feed his chicks. A large territory also helps him avoid his offspring being eaten: the bigger your territory, the less likely your particular nest will fall victim to a predator.

How big is a typical territory?

There is no such thing as a 'typical' territory! A colonial nesting species such as the Guillemot nests within pecking distance of its rivals; while large birds of prey such as Golden and Bald Eagles may defend an area of several hundred square kilometres.

Bald Eagles build huge nests from sticks, adding new material each year

What determines the size of a bird's territory?

It used to be thought that the size of a territory increased relative to the size of the bird, but although passerines do tend to have fairly small territories (anything from one-10th of an acre to more than 10 acres), so do colonial nesting species such as Gannets. In fact the size of a territory is generally determined by the food supply: seabirds nesting on a rocky island can be close together because they have an abundant and accessible supply of fish on their doorstep. In a sense, a large colony like this occupies a collective territory.

A Gannet colony on Bonaventure Island, Canada

How do birds defend a territory?

Strategies for protecting your patch vary from one bird to another. Seabirds use ritualised movements of the head and bill to warn off rivals; and if they come too close for comfort, a swift peck usually gets the message across! Birds with larger territories, such as raptors, patrol them regularly; to make sure that a rival has not sneaked into their area. But by far the most common method of defending a territory is singing: more than half the world's birds are songbirds, and use sound in order both to defend a territory and attract a mate (see Chapter 7).

A pair of Mute Swans tending their nest

When do birds pair up?

Most species, especially songbirds, pair up on the male's breeding territory as a direct result of the male's song or display. However, some birds choose a mate well in advance, while still on their wintering grounds. This particularly applies to wildfowl, with ducks, geese and swans displaying and pair bonding well before the breeding season; often thousands of miles away from where they will eventually nest. The pair will then travel together to the breeding grounds, where they can begin the process of nesting immediately, without wasting valuable time on the preliminaries.

Do females play hard to get?

Yes, very much so. As with humans, courtship amongst birds is a complex series of games and strategies. At the start, the female often rejects the male's advances, so that he becomes even more enthusiastic in his approach. Gradually she becomes more responsive, until she finally allows him to copulate with her and the pair is formed. She appears to be testing his fidelity and persistence: if he has to make such an effort to win her, perhaps he is less likely to abandon the nest and young.

Do females ever display to males?

Yes. In some species the female not only has brighter plumage than the male, but also takes the lead in courtship, leaving the male to incubate the eggs and raise the young. Such sexually liberated species include all three phalaropes, the Painted Snipe and Dotterel.

Who finally chooses a mate – the male or female?

It used to be assumed that the male was in the driving seat from a courtship point of view – after all, he usually takes the lead in courtship display. However, studies have since revealed that the female often makes the final choice of whether to accept or reject a suitor, so in reality she is in charge.

Unusually amongst birds, the female Dotterel takes the lead in courtship

Do birds fight each other over a mate?

Some species get into serious squabbles. European Robins are particularly aggressive, with males sometimes even fighting to the death over territory or females.

How long do males and females stay together?

That depends. In fact almost every kind of relationship between the sexes occurs among birds. Thus a male and female may pair for life, as in albatrosses and swans, or meet only a handful of times at a 'lek' (see page 165), as in some species of grouse and manakin. Other birds stay together for several days, weeks or months; and in some species males mate with several females, or the other way around. Despite these anomalies, however, the vast majority of birds form pairs for a whole breeding season or longer.

Do any birds pair for life?

Mute Swans usually pair for life

A few do; the best-known example being Mute Swan, in which it usually takes the death of one of the pair to split up the partnership. In most species, however, the pair bond usually lasts for the length of the breeding season, and even then may be disrupted by infidelity or the taking of multiple partners. Pairing for life is usually found in birds that live for a very long time, such as albatrosses and large raptors.

Which species has the most bizarre courtship display?

There are many candidates for this honour. The shortlist would have to include the Guianan Cock-of-the-rock from South America, in which the male displays his fabulous plumage to several females in a 'court' inside a forest clearing; the larger albatrosses, which face each other and point their bills at the sky while making extraordinary noises; and the bowerbirds of Australasia, the males of which build ornate 'bowers' decorated with all kinds of colourful objects in order to attract a female. There is also the famous 'penguin dance' of the Great Crested Grebe, whose males and females wave weed in each other's faces while standing up in the water using their feet as paddles. These bizarre performances have one thing in common: they all work. Otherwise, of course, the performers would have chosen a different strategy long ago.

Opposite: The famous 'penguin dance' performed by Great Crested Grebes

Do some birds have more than one mate?

Yes, quite a number of bird species regularly 'play the field'. A small minority are habitually polygamous – i.e. a male mates with more than one female (polygyny), or a female mates with more than one male (polyandry). Polygyny is more common, occurring in about two per cent of the world's bird species – primarily songbirds such as the Corn Bunting, where males may mate with up to seven females in a single breeding season. Polyandry is much rarer, occurring mainly in groups such as the phalaropes, in which the female is also more colourful and dominant than the male. A very few species indulge in 'polygynandry', where several males mate with several females – the avian equivalent of 'swinging'. This is practised by various ratites, such as Ostriches, rheas and the Emu, and even by our own humble Dunnock, as revealed by Sir David Attenborough in the BBC TV series *The Life of Birds*.

Can birds be unfaithful to their partner?

Yes, definitely. Although relatively few species are genuinely polygamous, many indulge in what might be described as 'a bit on the side'. A male does this in order to increase his total number of young; with the added advantage that chicks being raised in another male's nest do not require any work on his part. The more chicks he manages to have, the more he furthers the continuation of his genetic line – which is really his principal purpose in life. A female may 'play away' in order to optimise the quality of her offspring (by not laying all her eggs in one genetic basket, as it were). The result is that many males are raising at least a proportion of young which are not their own – whether they know it or not. Recent studies have revealed that almost half the chicks in an Indigo Bunting's nest may not be the offspring of the resident male.

Australian Emus have a complex sex life, with multiple partners

So why do any birds bother with monogamy?

Because it works. Monogamy provides the female with a partner to share the hard labour of raising offspring: by taking his turn to incubate the eggs, feeding her while she is incubating, and bringing food to the nest after the young have hatched. It also gives the male his best shot at raising a successful brood of young, and means he does not have to waste time and energy seeking out several prospective mates. It has been estimated that about 90 per cent of the world's birds are essentially monogamous (though at least some of these still grab other mating opportunities if they present themselves).

Male Black Grouse perform at a collective display-ground known as a 'lek'

Size does matter

The longest bird penis belongs to the Lake Duck, a fact discovered by scientists from the University of Alaska. It measures 42.5cm (17in) long: about the same length as the bird itself. The organ is corkscrew-shaped, allowing it to retract into the bird's body when not in use. It is thought that this extraordinary length is due to sperm competition driven by forced copulation, leading male anatomy to ever-greater extremes. Male Lake Ducks are known to be very promiscuous, and have been described as 'boisterous in their sexual activity'.

What is a 'lek'?

A lek is a communal area in which males gather to perform public displays in order to attract watching females, who then choose the most impressive male. The term comes from a Swedish word meaning 'to play'. Lekking behaviour is a bizarre adaptation, because once a male has successfully mated with a female their paths will not cross again – thus leaving her to do all the work in raising a family. It has evolved in several parts of the world, and amongst a wide range of different families, including grouse, manakins, waders such as the Great Snipe and Ruff, some species of hummingbird, and several different birds-of-paradise. It appears to occur only where there is a plentiful food supply, which means males do not have to defend territories in order to provide enough food for their offspring.

Is lekking unique to birds?

No. It is also practised by several species of mammal, including antelopes, walruses and fruit bats.

How do birds actually mate?

As in humans, birds mate by engaging in a sexual act known as copulation. Normally the male mounts the female, and they manoeuvre around until the male's cloaca (a single opening beneath the tail which is used both for excretion and reproduction) engages with that of the female. Sperm is then passed from the male into the female, where it fertilises her eggs. Incidentally, birds' sexual organs shrink dramatically outside the breeding season, in order to reduce their weight for flying.

Where do they mate?

Most birds mate either on the ground or while perching, as this maximises the chances of success. However, many waterbirds such as ducks will copulate in the water, while a few, such as swifts, even manage to do it on the wing.

Are any birds gay?

Not in the strict human sense of engaging solely in sexual relationships with a partner of the same sex. However, as with mammals, male birds will sometimes mount another male as if to copulate. This may be a case of mistaken identity (especially in species where males and females resemble each other); an act of aggression; or simply an outburst of sexual frustration.

It occurs most often amongst colonial nesters, especially gulls, where there may be an imbalance between the sexes.

A hybrid Ruddy x Common Shelduck (left)

Do birds ever mate with others from a different species?

Yes, frequently, although such pairings do not always produce offspring. 'Hybridisation', as it is known, generally occurs between individuals of two closely related species, whose genetic make-up is similar to one another's. It is particularly common amongst certain groups, especially ducks and geese, amongst which more than 400 different combinations have been observed, including Pochard x Tufted Duck, Mallard x Shoveler, and Goldeneye x Smew – the resultant offspring causing much confusion amongst birders.

Do hybrid offspring look more like one parent than the other?

Sometimes, yes. For example, any hybrids involving a Canada Goose tend to show the distinctive dark head and pale face patch of that species. Otherwise they often show an equal mixture of characteristics, enabling an observer to work out their origin. In a few rare cases a hybrid will show a plumage feature that does not appear to come from either parent, the result of an ancestral characteristic revealing itself.

What becomes of the hybrid young?

The offspring of such pairings are usually infertile, and therefore do not last more than a single generation. However,

some hybrids are not only fertile, but are more fertile than their parents. This often occurs in captivity, where poultry or cagebirds are deliberately crossbred to create newer, 'better', strains. It can also happen in the wild, as when various species of Darwin's finches on the Galapagos hybridised during unusual weather conditions brought about by 'El Niño'. These hybrids had larger bills, which proved better suited to the new conditions, and as a result thrived in competition with their parent species. Scientists are still puzzling over the implications of this extraordinary event on our definition of what makes a species.

A hybrid goose, one of whose parents is a Canada Goose

NESTS

What is a nest?

In its broadest sense, a nest is simply a place where an egg or eggs are laid, the young hatch, and (in some species) where the chicks are brooded until they are ready to leave.

Like many herons, Tricolored Herons nest in trees

Is nest-building unique to birds?

No. Although birds are by far the best-known nest-builders, other groups of animals also do so. These include the King Cobra, which drags dead vegetation into a small heap by using the coils in its body; several apes, notably the Orang-utan, which constructs large 'nests' for sleeping and shelter from the rain; and a whole range of fish, spiders, turtles and even crocodiles. Perhaps the best-known nest-builders other than birds were the dinosaurs.

Why build a nest?

All nest-building creatures do so for the same reasons: to protect the eggs and/or offspring from the elements; to safeguard them against predators; and to create a convenient single location in which to tackle the tricky work of incubating eggs and feeding young.

Do all birds have a nest?

No. The Fairy Tern precariously balances its egg on a horizontal branch, while the male Emperor Penguin incubates its single egg on the upper side of its feet, keeping it warm by using a special fold of skin. The alternative would be to lay its egg on the surface of the ice, where it would rapidly freeze.

Fairy Terns lay their single egg straight onto a dip on a horizontal branch

Building big

The largest nest made by any bird is that of one of the megapodes, the Orange-footed Scrubfowl of Australia. One breeding mound was estimated to measure 18 metres long, five metres wide and three metres high (roughly 60 x 16 x 10 feet), and may have weighed as much as 50 tonnes. The largest tree nest is that of the Bald Eagle. Enlarged annually, one nest in Florida measured six metres deep by three metres across (20 x 10 feet), and weighed almost three tonnes – the equivalent weight of two army jeeps. Of communal nesters, the African Sociable Weaver builds a colonial nest measuring up to eight by two metres (26 x 6 feet), with more than 100 individual chambers.

Gentoo Penguin with its chick

Grey Herons nest in colonies known as 'heronries'

How many different kinds of nest are there?

There are about a dozen different basic designs, on which each species fashions its own particular variation. The simplest is just a scrape in the ground, as favoured by some waders; or a natural depression in a rock, used by many colonial seabirds. This is hardly a nest at all in the traditional sense, since it involves no construction, yet to the bird itself it is just as important as the most ornate edifice built by other species. A more typical nest is the cup or bowl-shape favoured by the majority of songbirds. This is usually made from grass or twigs, and lined with vegetation or mud – or a variety of other materials including lichen, spiders' webs and even saliva. This highly versatile design is used by many species, from the world's smallest hummingbirds to the Bald Eagle. The next most popular design, used by everything from woodpeckers to owls and tits to trogons, is in a hole: either a cavity in a tree or a burrow made in earth or sand.

The nest of the Glossy Swiftlet is used for making bird's nest soup

Magpies build untidy nests out of sticks

What about more unusual nests?

Other, less conventional designs include floating nests (e.g. grebes, coots); nests stuck to the sides of buildings (e.g. House Martin); and the giant mounds built by the megapodes – gamebirds that pile up decomposing vegetable matter in order to incubate their eggs. More complex structures include domed nests, such as those of the Long-tailed Tit, Wren and Magpie, or those created by various tropical songbirds in order to protect their eggs and chicks from the sun – some of which, including that of the Cape Penduline Tit, even incorporate a false entrance in order to confuse predators. The most complex nest of all is the enormous multi-chambered structure created by the Sociable Weaver, from the Kalahari region of southern Africa, which may contain as many as 100 individual nests.

Do both sexes build the nest?

This varies from one species to another. In the majority of species the female does most of the nest building, though there are many species in which the male and female share the work equally. In some species the male does more: for example, the male Wren may build the beginnings of up to eight different 'cock's nests', which are then inspected carefully by the fussy female until she is satisfied with a particular one and completes it.

How long does nest building take?

That depends on the complexity of the structure. Some nests, such as those favoured by colonial seabirds, barely require 'building' at all; while the most sizeable structures such as those made by the megapodes may take several weeks. Most songbirds typically take between three and nine days: not bad, when you consider that, for example, a Long-tailed Tit's nest may contain up to 2,300 feathers.

Are some nests better built than others?

In the case of the typical cup-shaped nest, the soundness of its structure is critical to the breeding success of the occupants. A badly built nest will probably not survive bad weather or an inquisitive predator, while a well-made one can stand up to almost anything. Experience is vital: older birds tend to be better builders, as they have learned the complex skills of weaving nest material into a strong structure. For them, practice really does make perfect.

How do birds remember where their nest is?

As with finding their way on migration, birds use a number of both innate and learned cues in order to relocate their nest site. What is extraordinary is how rigid this process can be. When scientists moved a gull's eggs a few feet outside the nesting scrape, the adult bird returned to the original nest site and proceeded to brood in the very same place, ignoring the eggs themselves though sometimes they do roll the egg back into the nest. This proved that it is fidelity to the site, rather than to the actual eggs, which matters.

Do birds build more than one nest?

As well as the habit of some birds – such as the wren – of building several nests in order to offer females a choice, it is quite common for a bird to abandon a half-completed or even finished nest and start again elsewhere.

American Marsh Wrens build a nest from grass and rushes

This usually happens when the adults get wise to something that might harm their breeding success, such as the presence of a predator, or disturbance by humans. Some species of weaver are known to build 'decoy nests' in order to fool predators such as snakes.

Do birds build a new nest every year?

Most do. Indeed even species that have two or three broods in a single season usually build a new nest for each one, though they may pinch material from the old nest in order to save time. Others, including many large eagles, continue to add to the old nest year after year until it becomes a truly enormous structure.

Why do some birds breed in colonies and others alone?

Nesting in a colony, together with several hundred (sometimes many thousands) of other birds, has several obvious advantages. The main one, as with flocking and communal roosting (see Chapter 5) is 'safety in numbers': by being part of a mass of birds, each individual dramatically reduces its chances of falling victim to a predator. This is particularly important for seabirds, where the young stay in the nest for several weeks or months and would be very easy to attack if the bird nested alone. The food supply around a colony is usually abundant, so that there is no advantage to be gained in nesting apart from other birds.

Puffins nest in loose colonies, digging burrows where their egg and chick can be safe

So why don't all birds nest in colonies?

As always, what suits one particular bird will not necessarily suit another. Colonial nesting has its downsides, including more competition for mates, a greater struggle for food and space, and a higher risk of disease. Ironically, colonies are also more visible to predators, though the risk to individuals may still not be as high as nesting alone. Colonial nesters such as Gannets also use up a lot of energy in skirmishes and occasional all-out fights with their neighbours.

How common is colonial nesting?

Not very: only about one in eight of the world's bird species nest in colonies. However, amongst some groups it is much more common: more than nine out of 10 species of seabird nest colonially.

Do some birds deliberately nest alongside other kinds of bird?

Yes. Apart from the obvious case of colonial species, some birds deliberately seek out another species when they choose a nest site. So Long-tailed Ducks often nest in Arctic Tern colonies, where they gain the advantage of protection against predators from these noisy and aggressive birds. Likewise waders such as Dunlins will nest near Redshanks or Lapwings, both of which are famed for the hassle they dish out to intruders. Most bizarre of all, geese may choose to nest alongside predators such as Peregrines or Snowy Owls, which presumably keeps them safe from other predators such as Arctic Foxes.

House Martins build their nest from mud under the eaves of our homes

Do different kinds of bird ever use each other's nests?

Yes, frequently. Large structures such as White Storks' nests often host smaller 'squatters' such as House and Tree Sparrows, while some species habitually take over the nests of other birds – such as Carrion Crows, which often usurp a Buzzard's nest, sometimes evicting the rightful owners in the process. But the most inviting nest to squatters appears to belong to a peculiar stork-like bird called the Hamerkop, found throughout sub-Saharan Africa. The Hamerkop builds a huge structure out of mud, sticks and other debris in a tree fork, and intruders – including Egyptian Geese, eagle owls and even monitor lizards – often move in before the building is even complete.

What about nesting near other wild creatures?

Some birds will seek a measure of protection by nesting alongside other wild creatures. So certain tropical songbirds, such as Blue Waxbills, nest alongside colonies of aggressive biting or stinging insects, presumably because this affords them protection against any predator trying to attack the nest. In Africa, the Water Dikkop (a close relative of the Stone-curlew) often nests near crocodiles, again as a form of protection. Finally, many species live close to (or even in the homes of) human beings, either to take advantage of ready-made nest-sites or possibly as some form of protection. Examples include the aptly named House Sparrow, House Martin and Barn Owl, as well as various members of the crow and pigeon families.

EGGS AND INCUBATION

Do all birds lay eggs?

Yes. Birds are the only class of vertebrate that *never* give birth to live young; although the majority of amphibians, fish and reptiles also lay eggs. Three primitive mammals, the Duck-billed Platypus and two species of echidna (known collectively as monotremes), lay eggs too, as did the dinosaurs.

Why do birds lay eggs?

The egg is one of nature's best ways of providing protection and food to a growing chick. But the question remains: why don't birds do as the vast majority of mammals, and keep the young safe in the mother's body until it is time to give birth? The reason is related to the fact that birds are generally lighter for their size than other creatures, due to their need to fly. So the extra weight of the young birds inside a mother's body would make it difficult – perhaps impossible – for her to get airborne, and increase the risk of being caught by predators. By laying eggs, usually in a nest, the female also enables (or obliges) the male to share in the process of successfully rearing a family.

How many eggs do birds lay?

This varies enormously from one family of birds to another. A few (many seabirds) lay just a single egg; others (hummingbirds, most birds of prey) lay two; many (most songbirds, waders etc.) lay between three and six; some (certain smaller songbirds, including tits) lay seven to 12; and a small minority (gamebirds such as pheasants, partridges and quails) lay a dozen or more. However, if eggs are continually removed from a nest, some birds will continue laying almost indefinitely. Hence the popularity of domesticated ducks and chickens.

Gamebirds such as Pheasant lay large clutches of eggs

Eggstraordinary!

The largest egg laid by any living bird is that of the Ostrich, whose egg measures an average of 16cm long and 13cm in diameter (six by five inches), and weighs about 1.5kg (3.3lbs) – roughly 24 times the size of an average hen's egg. It takes about 45 minutes to hard-boil one.

The largest egg ever laid was that of the extinct Elephant Bird of Madagascar, whose egg weighed up to 12kg (26.5lbs) and measured 38 x 30cm (15 x 12in) – about eight times the size of an Ostrich egg. It is also thought by some scientists to have been larger than any dinosaur egg.

The largest egg relative to body size is laid by the Little Spotted Kiwi of New Zealand, whose egg weighs about 250g (just under 10oz), almost one-quarter of the body weight of the female, and the equivalent of an ostrich egg weighing 25kg (55 lbs).

The smallest egg is that laid by the world's smallest bird, the Bee Hummingbird. Its egg measures about 12.5mm long by 8.5 mm in diameter (0.5 x 0.33 in) and weighs just half a gram – about half the weight of a paper clip. It would take 125 Bee Hummingbird eggs to equal the weight of a hen's egg, and an incredible 3,000 to equal the weight of an Ostrich's. The smallest egg relative to body weight is that laid by the Emperor Penguin, which at about 500g (1.2lbs) is just 1.5 per cent of the adult's body weight of 30–40kg (66–88lbs); and the Ostrich, whose egg also weighs roughly 1.5 per cent of the adult's body weight.

An Ostrich egg next to a Chicken's egg

A female Ostrich settles down on her huge clutch of equally huge eggs

Why do some birds lay more than others?

This depends mainly on two factors: the likely survival rate of the young, and how much effort it takes to raise them. So large, long-lived colonial nesters such as the Wandering Albatross lay only a single egg, and devote all their efforts to raising the chick over a very long period (about nine months from hatching to fledging). Songbirds, which have a very high death rate and a short lifespan, tend to lay large clutches to maximise the chances of one or two chicks reaching adulthood. It is interesting to note that resident species such as the Blue Tit tend to have larger clutches than some migratory species such as the Whitethroat.

How soon after fertilisation is an egg laid?

The actual process following fertilisation, which includes the deposition of the albumen, the formation of the membranes and the laying down and colouration of the eggshell, takes roughly 24 hours for most species.

Is a clutch of eggs all laid at once?

The most common pattern of egg laying is one egg every 24 hours, found amongst songbirds, most ducks and geese, and smaller waders. Birds of prey, ostriches and larger shorebirds usually lay their eggs two or three days apart, while some seabirds have an even longer gap – up to a week in the case of the Masked Booby.

177

Are eggs laid at a particular time of day?

Many birds, including songbirds and hummingbirds, lay their eggs at dawn. This means they can immediately feed and gain the energy necessary to lay the next one. Pigeons and pheasants, however, tend to lay their eggs in the evening.

Are all eggs the same shape?

No. In fact, some are not even egg-shaped. The eggs of owls and kingfishers are almost round, while those of grebes and divers are long and thin. Guillemots' eggs are 'pear-shaped' (rounded at one end and tapered at the other), which reduces their chances of falling off narrow cliff ledges since they tend to spin in a tight circle when knocked. Many waders also lay pear-shaped eggs, as these fit snugly together with pointed ends turned inwardswhen laid in a clutch of four, making them easier to incubate.

Do large birds lay big eggs and small birds lay small ones?

In actual terms, yes; but not in relative terms. In fact, as a general rule, egg weight as a proportion of the female's body weight declines the larger a bird is, so that whereas the eggs of most songbirds each take up about 10–20 per cent of the female's body weight, those of the Ostrich take up less than two per cent of hers.

Why are some eggs white, while others are patterned or coloured?

In general, eggs are coloured or patterned to help disguise them from predators, and white when laid in holes, where predators are unlikely to

The world's smallest birds' eggs are those laid by hummingbirds – just a few millimetres long

Little Ringed Plovers nest on bare shingle or gravel, where their eggs are camouflaged from predators

be hunting by sight. White eggs may also be easier for the parent birds to see in the dark. It is thought that birds' eggs were originally white (like those of their reptile ancestors) and later developed colours and patterns as the need for concealment increasingly began to arise. Common patterns include spotting, marbling, streaking and blotching.

How long does incubation take?

This varies a lot, from just 10 days in the case of the Red-billed Quelea of Africa, to almost three months in the case of the larger albatrosses and the Brown Kiwi of New Zealand. Most songbirds incubate for between two and three weeks; while ducks and waders do so for slightly longer (three to four weeks); and raptors, geese and swans longer still (four to six weeks).

Do male birds incubate?

Yes: in many species both males and females incubate the eggs. However, in some only the females incubate and in a small minority of species the males do it alone, with no direct help from the female. These include various species of jacana and phalarope, and the Emperor Penguin of Antarctica.

Why do birds abandon their eggs?

Any bird may abandon its nest and eggs for a number of reasons, including predation, bad weather or disturbance. Better safe than sorry – after all, there's always next year. However, the

Emperor Penguins take it in turns to guard their single chick while the other parent walks long distances to get food

megapodes of South-east Asia and Australasia have made desertion a way of life. They build enormous mounds in which they bury their eggs, relying on the heat generated by the rotting vegetation to incubate them. Young megapodes hatch fully feathered and can look after themselves immediately, never knowingly meeting their parents. The same, of course, is true of young brood parasites such as cuckoos (see page 181).

Long-eared Owl chicks in their nest

Clutch control

The largest clutch laid was that of a female Grey Partridge, which was reliably recorded as comprising 19 eggs. Clutches of up to 28 eggs have been recorded for several species of gamebird, including the California and Bobwhite Quails of North America, but these are almost certainly the result of two or more females laying in the same nest. Amongst songbirds, Blue Tits have been known to have clutches of up to 19 eggs, possibly all from a single female. The most eggs laid in a single nest are those of gamebirds and the Ostrich, where several females often lay in the same nest. One Ostrich nest in Nairobi National Park, Kenya, contained a total of 78 eggs, though only 21 of these were actually incubated.

What is an 'addled' egg?

One that has failed to hatch, because the embryo has died.

What is an 'infertile' egg?

One where the egg was never fertilised, so does not contain an embryo. Infertile eggs are sometimes laid by unmated females.

Do eggs all hatch at the same time?

That depends. In many species, especially songbirds, incubation does not begin until there is a full clutch, so most eggs will hatch on the same day. Other birds, such as raptors, take a different approach: of two Golden Eagle eggs, one will generally hatch a couple of days before the other, meaning that the elder of the two chicks is generally larger and healthier. In a poor year for hunting, parents will mainly feed the larger chick, which can lead to siblicide. In a year when food is abundant both chicks will survive. Although this seems cruel, it is a sensible strategy: better one strong chick should survive than both die of starvation.

Do birds ever lay their eggs in other birds' nests?

Yes, frequently. When population density is high (as in colonial species), this practice is especially common. The interloper can thus increase the number of offspring to which it passes on its genes, without having to go through the tedious business of actually raising them.

What is a 'brood parasite'?

Technically, any species that lays an egg in another bird's nest. However, the term is usually reserved for those habitual cheats, such as several species of cuckoo and cowbird, which always lay their eggs in the nests of a different species, letting the host parents do all the work raising the young. This enables the female brood parasite to lay far more eggs than would normally be possible (up to 40 in a single breeding season), and maximises her chances of producing plenty of offspring.

A huge Cuckoo chick being fed by its foster-parent, a Redstart

How do the eggs fool the host species?

The innate urge to incubate eggs is a very strong one, which means that the host birds are already inclined to do so. However, the intruder may improve the odds for its own eggs by making them as close as possible in shape, colour and size to that of its host. This can lead to an evolutionary 'arms race' in which, as the hosts get better and better at detecting the intruder's egg, only those eggs closest in appearance to their own survive, becoming even harder to detect than before.

Why don't hosts reject chicks that are not their own?

Once the chick is hatched, the stimulus of an open mouth requiring food appears to be enough to fool almost any parent. This persists even when the young bird has grown to more than twice the size of the adult host (as in the case of the Cuckoo and the Dunnock).

How many brood parasites are there?

About one per cent of the world's birds (roughly 100 species) are brood parasites. This unusual phenomenon occurs in six unrelated families: the majority of Old World cuckoos and three species of New World cuckoos, five out of the six species of cowbird, most of the honeyguides of Africa and Asia, some African whydahs and widowbirds, indigobirds, and a single species of wildfowl – the Black-headed Duck of South America. This wide range of families shows that the habit must have evolved independently several times.

Sitting tight

The longest incubation period belongs to the Wandering and Royal Albatrosses (up to 85 days, but not always continuously) and the Brown Kiwi (also up to 85 days), while the longest continuous stretch of incubation is that of the male Emperor Penguin, who keeps a single egg warm by resting it on his feet for up to 67 days without a break. Amongst passerines, the lyrebirds incubate their single egg the longest, for about 50 days. The shortest incubation period is probably that of the Red-billed Quelea (the world's most numerous bird), which can be as short as 10 days.

Brown Kiwi

HATCHING, BROODING AND FLEDGING

How does a chick hatch?

As the embryo grows, it develops a sharp projection known as an 'egg tooth' on its bill, and a 'hatching muscle' at the back of its skull; while at the same time the eggshell becomes weaker, and an air space develops at the blunt end of the egg. Once the chick is ready to emerge, it pushes itself up to the blunt end, cracking the shell with its egg tooth. Then it uses its legs to move itself around the egg, producing tiny cracks in the shell. Finally it pushes itself out and emerges into the world, exhausted but free. This process usually takes a few hours, but may take anything from 30 minutes (most songbirds) to several days (albatrosses). The 'egg tooth' falls off after a few days.

What is 'brooding'?

The sitting on chicks by one or both parent birds, in order to keep them warm and safe against predators. The term is also sometimes used as a synonym for incubation, i.e. sitting on eggs.

A duckling hatching out of its egg

What is 'fledging'?

This is the process by which a baby bird becomes ready to leave the nest and fly. It takes place in several stages: while the youngster is in the nest it is called a nestling; when it leaves the nest but cannot yet fly and is still being fed by its parents, it is a fledgling; and it is only 'fully fledged' when it has shed its down, acquired a full set of feathers and flown the nest. The term 'fledging' applies mainly to birds such as songbirds, where the young are born helpless and stay in the nest for a period before they acquire their first full plumage.

Young raptors often leave the nest before they are fully fledged

Why do some birds leave the nest more quickly than others?

Many kinds of birds have young which are much more independent after hatching. They are able to leave the nest almost straight away, and can walk or swim, and in many cases find food for themselves. These are known as 'precocial' species, and include all wildfowl, waders and gamebirds. Most of these birds nest on the ground, where being able to look after oneself is an urgent survival priority. Other

An easy lay

The most eggs produced during a single breeding season is 40, by the female Brown-headed Cowbird of North America. The reason for this impressive productivity is that the cowbird is a brood parasite, laying her eggs in other birds' nests, so does not have to incubate eggs or raise the young.

The Brown-headed Cowbird of North America

groups, such as gulls, are 'semi-precocial', with the young active but remaining in the nest and still being fed by their parents; while herons and owls are less self-reliant and are known as 'semi-altricial' species. All songbirds, in contrast, are 'altricial': they are born naked and blind, and must be fed by their parents until they are fully fledged (see opposite). Another term for precocial is 'nidifugous', while the corresponding term for altricial is 'nidicolous'.

Mallard ducklings are able to leave the nest and swim almost immediately after hatching

What should I do if I find a baby bird fallen out of its nest?

Leave it alone! The parents are almost certainly close by, and able to take care of it themselves. If you take it into your home it is highly likely to die. The only exception to this rule is if you find a naked, helpless nestling which has just fallen out of its nest; in which case it is worth putting it straight back and hoping for the best. Try to handle it as little as possible.

Which birds feed their young for longest after fledging?

Some seabirds, including frigatebirds, may feed their young for as long as 18 months after they fledge. They do so by stealing food from other seabirds and then regurgitating it for their offspring.

Which bird is the most sociable nester?

The aptly-named Sociable Weaver of Africa, which builds an enormous collective nest with up to 100 separate chambers in order to insulate the eggs and young against freezing winter nights in the Kalahari desert. The largest nests are over a metre high and several metres in diameter, and may house more than 300 birds at any one time.

How do birds keep their nest clean?

Many birds are very hygienic, taking away their offspring's droppings, which come enclosed in 'faecal sacs' (shrink-wrapped poo, if you like) to avoid soiling the nest. Others are less than house-proud, allowing the faeces to

A parent frigatebird tends its chick

A young Bald Eagle tests its wings

build up, which in tropical climates can cause quite a stink. Arguably the dirtiest nest of all are the hornbills', in which the female uses a mixture of mud and her own droppings to wall herself into the nest cavity (sometimes with help from her partner), and stays there throughout the incubation and brooding period while the male passes food through a small opening. She finally emerges in a very filthy state indeed.

How do young birds learn to fly?

Flight is an instinctive rather than a learned skill. So a baby Blue Tit leaving a nestbox will, after some hesitation, launch itself out of its nest-hole and into the air. While first flights are not always very successful or stylish, the birds soon

Leaving home

The longest fledging period is that of the King Penguin, at up to 13 months. Of flying birds, the Wandering Albatross takes up to 280 days (just over nine months). The shortest fledging period is nine days, found in several passerines including the Corn Bunting.

get the hang of it. Other birds, such as young Peregrines, will practise flapping their wings for several days before leaving the nest, which helps build up their wing muscles. Once airborne, they will also get 'tuition' from their parents – the first step on the way to becoming supreme masters of the air.

9 • WHERE DO BIRDS GO?
Migration

What is 'migration'?

Migration, at its simplest level, is defined as a regular seasonal movement of a population of organisms. Birds are probably the best known migrants, but it is also a way of life for an amazing variety of other animals – from anchovies to zebras. Usually, though not always, migration takes place between breeding and wintering areas; so usually, though not always, it involves two journeys each year: one outward, and one return.

Why do birds migrate?

Some birds migrate in order to find food and somewhere to breed, which may not be available in one place all year round. Take a typical insect-eating European migrant such as the Willow Warbler. After breeding in Britain and northern Europe during summer, it flies up to 10,000km (6,200 miles) to southern Africa in order to avoid the cold, foodless northern winter. There, it spends the winter (the southern African summer) in a warm, sunny environment with plenty of insects to eat, before returning north in spring when conditions are once again suitable for breeding.

Willow Warbler

Isn't migrating riskier than staying put?

Although undertaking such a long journey may seem a risky strategy, the dangers of staying put for the winter, in a cold climate with little or no insect food, far outweigh the dangers encountered on migration. Migration

Goldcrests may be small, but they are tough – able to survive very cold winters in northern Europe

gives birds the best of both worlds. In fact, perhaps we should turn the previous question on its head and ask 'why don't all birds migrate?'

So why do some insect-eating birds spend winter in Europe?

A few mainly insect-eating species, including the Goldcrest and several species of tit, do remain in northern and western Europe all year round. They survive by exploiting hidden food sources, such as tiny insects hiding beneath the bark of trees. It is lack of food – not cold weather itself – that kills birds in winter. So even in Arctic Norway, Goldcrests are able to survive in temperatures of more than 30°C below freezing, because they can find plenty of food deep in the heart of coniferous forests.

So why do migrants bother coming back again?

Most songbird migrants from Europe and North America spend the winter in sub-Saharan Africa and South or Central America respectively. But the abundant food supplies they find there do not last all year round. If they were to remain through to the breeding season, when extra energy is required to produce eggs and feed young, it might become harder to find enough – especially when there is so much competition from local, resident species. Birds that return to temperate areas such as Britain or the United States and Canada tend to have larger clutches (and sometimes more broods) than birds in tropical regions.

This all suggests that the plentiful food supply available in the northern summer allows these birds to enjoy a more successful breeding season than if they had remained in the tropics. The longer daylight hours of the northern summer also assist with the rearing and rapid growth of young birds, as there is more time to search for food.

Do all birds from one species migrate?

No. Some species are partial migrants, with different populations following different strategies. For example Robins from Scandinavia migrate to Britain for the winter, while many British Robins stay put. Alternatively, males and females of the same species

may go their separate ways: female Chaffinches leave Sweden in the winter, while males stay on or near their breeding areas. This led the scientist Linnaeus to give the species the scientific name *Fringilla coelebs*, meaning 'bachelor finch'.

How did migration come about in the first place?

It is often said that migrating birds head south for the winter – but this puts the cart before the horse. In fact it is thought that many migratory birds originally evolved in equatorial regions, and first headed north to avoid competition with other species there. By doing so, they could also take advantage of the long daylight hours and plentiful food supply of the northern summer. But they still had to

In winter the Robin in your garden may have travelled across the North Sea from Scandinavia

return south again every winter, when the weather closed in. Thus migration was the key that opened up new lands for birds. It soon became a viable way of life for many species.

Why do birds go so far?

Surely migrants heading north in spring would be better off stopping to breed around the Mediterranean, than pressing on to the Arctic? Well, many of them do. But those that do travel to the Arctic Circle gain the advantage of even longer hours of daylight and less competition from other species. This explains why birds such as the Red Knot undertake a huge journey from the southern hemisphere to the edge of the Arctic to breed – food is abundant there, and far fewer species are competing for it.

What proportion of the world's birds migrate?

Approximately 4,000 species – roughly 40 per cent of the global total – are usually considered to be migratory, though not all undertake long-distance journeys.

How many individual birds migrate?

Taking the species that undertake major north–south journeys in the Old World, up to five billion birds of more than 200 species travel south to Africa each autumn. Of North America's 20 billion or so birds, it is thought that about one-third – more than six billion individuals – migrate at least as far as Central America.

Red Knots are long-distance global migrants

Before it migrates, a Sedge Warbler will build up fat reserves for the long journey ahead

during this period. Other birds, especially waders, will put on some fat beforehand, but also stop to feed on the journey south, replenishing their energy resources as they go. However, large birds of prey such as the Osprey are unable to fatten up for migration, as this would make them too heavy to fly long distances, so they stop off to feed at regular intervals en route.

So what makes them set off?

The main impulse for departure both from winter quarters and breeding grounds is tiny changes in day-length, which affect the bird's brain. The brain triggers the bird's endocrine system to produce hormones that stimulate it to prepare for the long journey ahead – for example, by feeding hard in order to build up fat reserves. Even cage birds may show signs of restlessness during spring and autumn, when migration is in the air.

How do birds prepare for migration?

Birds prepare for migration in two ways. Most moult their worn adult feathers or first juvenile plumage, acquiring a spanking new set of feathers in preparation for the journey. They also build up fat reserves, which may take several weeks, as they feed frantically in order to increase their weight by up to 50 per cent. The fat is stored in a layer just beneath the skin, and can be seen when migrating songbirds are caught for ringing. A Sedge Warbler may increase its normal weight from 10g (0.4oz) to as much as 15g (0.6oz)

Does the weather affect departure times?

Yes, especially in autumn, weather conditions can play a big part in the timing of the birds' departure south. While the birds may be physically ready to leave, they need to use the best weather conditions for migration, so may stay put if the weather is bad. Thus songbirds leaving Scandinavia or eastern Canada will usually wait for the passing of a cold front with clear skies and following winds that help them on their way.

Air miles

The world's longest migratory journey is undertaken by the Arctic Tern, from its sub-Arctic breeding grounds to and from the Antarctic – an annual round-trip of up to 35,000km (approx. 22,000 miles). During its lifetime, a single bird may travel more than 800,000km (half a million miles) – more than any other species. As a result, the Arctic Tern experiences more daylight than any other living creature. One individual, ringed as a chick near Murmansk in Russia, was recaptured alive a year later near Fremantle in Western Australia – more than 22,400km (14,000 miles) away. The longest journeys of land birds include those made by the Barn Swallow, which migrates in the Old World between Norway and South Africa, and in the New World from Alaska to Argentina – both journeys of roughly 9,600km (6,000 miles). But this is dwarfed by the journey of the White-throated Needletail, a species of swift, which undertakes a twice-yearly trip between Siberia and Tasmania of roughly 12,000km (7,500 miles) each way.

The Arctic Tern has the longest migration of any of the world's birds

Do birds know when to migrate?

The notion that birds somehow 'know' when to migrate goes back at least as far as the Old Testament, where it is written that "the turtle-dove and the swallow and the crane observe the time of their coming". Especially in spring, it does seem that some species arrive on pre-set dates – as can be seen from the widespread folklore on the arrival times of familiar species such as the Barn Swallow and Cuckoo. But birds don't follow a strict calendar – at least not in the sense that we understand it. Instead they respond automatically to certain natural stimuli.

Turtle Dove migration is mentioned in the Old Testament

What about climate change?

In recent years, global climate change – and its impact on weather patterns and temperatures – means that some birds appear to be arriving earlier than usual in spring and staying later in autumn. It is too early to say yet whether or not this will permanently change the timing of bird migrations. Climate change is also causing birds problems by reducing their habitat: for example, drought in the Sahel Zone of western Africa has led to the desertification of large areas, depriving migrants such as Sand Martins of the vital green areas that sustain them on their long haul journeys. The climatic phenomenon known as 'El Niño' is also having a major effect on both breeding and migrating birds, by dramatically changing the world's weather patterns, especially in the tropics.

Do all migrants arrive and depart at the same time?

No. Different species arrive and depart at very different times on both outward and return migrations. In northern Europe, for instance, two of the earliest spring migrants are Wheatear and Sand Martin, both of which generally arrive in March; while late arrivals include Turtle Dove and Spotted Flycatcher, which do not usually return until mid-May. Most Cuckoos and Swifts have gone by August, while Swallows may stay on well into October. There are regional differences too: the farther north you are, the later the migrants arrive in spring, and the earlier they depart in autumn.

Migrants use many navigation aids, including the moon, the stars and the Earth's magnetic field

How do migrating birds find their way?

This is one of the greatest of all natural mysteries. There is no one, single method used by birds to find their way; instead, most pick and choose from a number of orientational tools. The most important are the earth's magnetic field, and visual compasses such as the sun (for daytime migrants), and the moon and stars (for night-time ones). Other useful tricks include the ability to perceive polarised light (especially when clouds obscure the sun), and 'vector navigation' – similar to the 'point-and-compass' method used by early sailors.

How do birds use the Earth's magnetic field?

It has long been known that birds possess some form of internal compass, which enables them to detect the earth's magnetic field and use it to orientate themselves

in the right direction. Scientists have discovered a substance called magnetite in the skull of pigeons, which must enable them to sense their position in this way. Experiments have also shown that birds respond to an artificial magnetic field, which can be used to confuse their normal senses.

Do birds migrate by night as well as day?

Yes, very much so. In fact more species – and many more individuals – are night flyers, and for several reasons. First, the air is cooler at night, which is especially important as the bird goes farther south. Cool air allows a bird to fly faster, with less dehydration and loss of energy. Second, there are fewer predators around, most of which are daytime migrants. Finally, migrating by night and stopping by day allows birds to feed during daylight hours. Nocturnal migrants include most songbirds, wildfowl and waders.

The Ruddy Turnstone is a passage migrant in Britain

What is a 'passage migrant'?

A passage migrant is a species that passes through a particular area in spring and autumn, but does not generally stay to breed or winter there. Black Tern and Little Gull are regular passage migrants to Britain; as are Curlew Sandpiper and Little Stint, two waders that breed in the Arctic and winter in West Africa. However, how you define a passage migrant depends on where you are standing at the time: the Ruddy Turnstone is a passage migrant throughout much of its range, but a breeding visitor to the Arctic and a non-breeding visitor to the coasts of Africa and South America.

How long do birds' migratory journeys take?

Many smaller migrants, such as the Red Knot and Sedge Warbler, work on the 'long-hop' system, covering hundreds of miles a day, and reaching their destination in just a few days. Others, including the Osprey, take their time, stopping off to feed for a few hours or even days, before restarting their journey south. Radio tracking of Ospreys has revealed that juveniles can take up to three months to travel from Scotland to their winter quarters in West Africa.

Do all migratory birds follow the same route?

As you might expect, different species follow very different routes. Raptors such as eagles and buzzards find it hard to fly over large expanses of water because of the lack of thermal air currents to give them lift. So most European birds of prey take routes south which involve the shortest possible sea crossing. They cross at the Strait of Gibraltar, over the Bosphorus at Istanbul, and over the Red Sea from Eilat in Israel, and can be seen in huge concentrations at these locations every spring and autumn.

Why do some birds migrate in V-formation?

Some larger birds, notably swans, geese and cranes, often travel in a V-formation. This is an energy-efficient way for them to cover large distances, as the leading bird creates uplift and reduces wind friction for the others (just like racing cyclists riding in a pack). The lead bird is usually an experienced adult, and will take turns with others to avoid becoming exhausted.

Geese migrate in V-formation to save energy

How high do migrating birds fly?

Different groups migrate at different altitudes: raptors at around 600–1,000 metres (2,000–3,300 feet), and most songbirds at below 1,500 metres (5,000 feet), while waders and geese often travel at 1,500–3,000 metres (5,000–10,000 feet). Occasionally they venture even higher: there are a few records of migrating wildfowl and raptors topping 6,000 metres (20,000 feet).

Why do some birds migrate by day?

Birds which can feed 'on the go', such as swallows and swifts, generally migrate by day, feeding on flying insects on the way. Other diurnal migrants include birds of prey, which because of their weight and soaring style of flight rely on thermals (rising currents of warm air) to gain altitude for migration.

Swallows mainly migrate by day

The Straits of Gibraltar are a regular crossing-point for millions of migrants between Europe and Africa

How do they know when they have reached their destination?

As birds near their final destination, visual landmarks such as coastlines, rivers and mountain ranges, wind direction, and sound and sometimes smell come into play; allowing them to return, in some cases, to the very place where they were born. Of course, such landmarks don't help when young birds are making the outward journey for the first time – which just demonstrates the additional importance of a genetically inherited mental map.

Do young and adult birds all travel together?

Not always. Although some groups of birds, notably cranes and wildfowl, often travel in family groups, the young of most other species are left to their own devices. Many adult waders and songbirds leave several weeks earlier than the juveniles, which must undertake their first journey without help. The most extreme example of this is the juvenile Eurasian Cuckoo, which is abandoned by its parents even before birth. Yet despite the lack of parental guidance, these species are no less successful than those where adults and young travel together. This shows that many orientational techniques are innate, rather than learned.

White Storks usually return to the same nest-site; some have been used for decades

Do birds coming back in spring follow their autumn route in reverse?

Most do, but a few species undertake what is known as loop migration. In North America, many species of wood-warbler travel across the western Atlantic Ocean in autumn, taking advantage of tailwinds to follow the shortest possible route south. But in spring, when there are no winds to help them on their way, they must follow a longer route along the eastern seaboard of the USA.

Do any birds migrate north for the winter?

In Africa and South America, quite a few species that breed in the southern hemisphere head towards the Equator outside the breeding season, but in Europe and North America only a handful of species actually head North. These include Water Pipit, some of which breed in the Alps and head northwest

What goes up…

The shortest migratory journeys are those made by altitudinal migrants such as the Mountain Quail and Clark's Nutcracker of North America, which simply descend from the mountaintops to sheltered valleys during the autumn, and return to the high tops in spring. Mountain Quails make the twice-yearly journey on foot.

Unlike most birds, Ross's Gulls fly north for the winter

to winter in Britain. Three species of seabirds breed in the south Atlantic during the southern summer (our winter), then head north to spend our summer in the northern Atlantic Ocean. These are the Great and Sooty Shearwater and the tiny Wilson's Storm-petrel. But perhaps the most extraordinary example of all is Ross's Gull, which breeds in Arctic Canada and Siberia, yet still heads north at the end of the breeding season to spend the winter on the edge of the Arctic pack-ice.

Do any birds head west for the winter?

Although the majority of migratory species are north–south migrants, there are many exceptions. Lots of waders and wildfowl breed in Siberia and Scandinavia, then head west or southwest in autumn, to take advantage of the much milder winter climate in Britain and Ireland. In doing so they are following the same principle as birds that migrate north to south: heading to a destination with milder weather and thus more available food.

What is 'altitudinal migration'?

As opposed to 'latitudinal migration', where birds head roughly south in autumn and north in spring, altitudinal migration occurs when birds breeding at high altitudes head down towards lowland areas for the winter. They do so in search of a milder climate and more ample food supplies; which often means wintering on or near the coast. Although these journeys may seem insignificant in terms of distance travelled, they often involve a major change of lifestyle.

Do all migrants make regular twice-annual journeys?

No – a few species are irruptive, meaning that they undertake occasional and unpredictable mass movements away from their breeding areas. These are normally a result of food shortages on their breeding grounds, sometimes combined with

The Two-barred (or White-winged) Crossbill is an irruptive species, making irregular migratory journeys

a population boom. Irruptive species may be rare or absent one year, and abundant the next. They include various species of crossbill, which generally migrate in mid-summer, and often stay put for a year or more, breeding in their new location before returning to their original one. Another famous example is the Bohemian Waxwing, which sometimes heads to Britain in large numbers from its breeding areas in Scandinavia and northern Russia.

Can birds change their migratory habits?

Yes, from time to time. A well-known example is the German population of Blackcaps, some of which within a few decades changed their winter quarters from Iberia and North Africa, to Britain and Ireland. This occurred as a result of a genetic mutation that sent birds in a northwesterly direction in autumn instead of a southwesterly one. Once on their new wintering grounds they found a mild climate and abundant food, enabling them to survive and return earlier to breed. As a result the mutant gene spread rapidly through the population, the whole of which now winters in Britain and Ireland.

Do migrating birds ever get lost?

Frequently, and for all sorts of reasons. In autumn, the vast majority of migrants are juveniles, on their very first journey south. Their lack of experience means they can easily become disoriented, especially when they hit bad weather. Wind and rain may prevent them from using visual cues such as the sun or stars, so they tend to drift off course.

What happens to lost birds?

Many end up exhausted and fall to the ground, or into the sea and drown; others allow themselves to be carried by crosswinds until they reach the safety of land. This gave rise to the theory of 'drift migration', developed to explain the regular arrival of eastern migrants on the east coast of Britain after easterly winds and rain. These birds generally reorient themselves and get back on course. But every autumn a few North American land birds get blown off course by westerly gales, cross the Atlantic and make landfall in Britain and Ireland, much to the delight of birdwatchers.

The Short-toed Eagle is a very rare visitor to Britain, so when one does arrive it usually attracts crowds of twitchers

What is a 'vagrant'?

In effect a vagrant is any wild bird found outside its normal range – either breeding, migratory or wintering – having arrived by natural means. Britain and Ireland play host to vagrants from all points of the compass, virtually all of which started out as migrants, but for one reason or another lost their way. There are several reasons for vagrancy, including unreliable orientation mechanisms, genetic mutations, and the effects of extreme weather conditions. Vagrancy is most common amongst juvenile birds undertaking their first journey, and so happens more often in autumn than spring, though it can occur at any time of year.

What is an 'accidental'?

Accidental is a synonym for vagrant. It used to apply strictly to a bird that had occurred fewer than 20 times outside its normal range, but the definition has since become more flexible, and can now mean any bird found outside its normal range.

The Great White Egret has recently colonised Britain as a breeding bird, having once been a rare vagrant

and various herons and egrets. When conditions are good, these birds may even stay on to breed.

Can good weather cause vagrancy?

Ironically, good weather can also lead to vagrancy. In spring, when high pressure brings fine, settled weather over western Europe, many returning migrants heading back to the Mediterranean region fly right over their intended target, and end up in Britain. This phenomenon is known as overshooting, and involves species such as Hoopoe, Bee-eater,

Apart from the weather, why else would a migrant go off course?

Other reasons for vagrancy include a kind of 'wanderlust', when young non-breeding birds simply 'go flyabout' and end up outside their normal range. This may have an evolutionary advantage, and eventually result in the species extending its breeding range.

Once a very rare vagrant, Pallas's Warbler is now an increasingly regular autumn visitor to Britain

Do vagrants ever find their way back home?

Not very often. Overshooting birds may well return south almost immediately, or spend the spring and summer in their new home before heading back to their winter quarters in the autumn. But for birds that have crossed the Atlantic, prevailing westerly winds make the return journey virtually impossible. Most songbirds perish soon after arriving, due to lack of suitable food, but some larger species, such as waders and gulls, survive on this side of the Atlantic for many years, possibly migrating back and forth between Europe and Africa, instead of between North and South America. Some may make a permanent home here: such as

'Albert', the Black-browed Albatross from the South Atlantic, who lived in a Gannet colony on the island of Unst in Shetland for many years from the 1970s to the 1990s.

Do birds ever go completely in the wrong direction?

Some birds appear to have a defective compass, and head off in entirely the wrong direction – a phenomenon known as 'reverse migration'. So each autumn Siberian species such as Richard's Pipits, and Pallas's and Yellow-browed Warblers, which should be spending the winter in South-east Asia, turn up in Britain. However, a new theory suggests that these birds are not in fact vagrants in the true sense

Snow Geese sometimes accompany White-fronted Geese on their migrations to this side of the Atlantic

of the word, but part of a pioneering population which may eventually establish itself as wintering in Western Europe.

What is a 'fall'?

A fall is the simultaneous arrival of large numbers of migrants (usually songbirds such as warblers, flycatchers, chats and thrushes), as a result of rough weather that forces them to make landfall.

What is 'abmigration'?

This occurs when an individual from one population or species accidentally joins a flock of another population or species, and migrates with the flock to their wintering areas. The following spring the disoriented bird usually stays with its new companions and therefore returns to a different breeding area. Abmigration often occurs amongst sociable migrants such as geese, where – for instance – a Snow Goose may join a flock of Greenland White-fronted Geese in western Greenland, and end up in Scotland instead of Texas.

What is a 'wreck'?

A wreck is a major displacement of seabirds, often as a result of autumn or winter gales. Onshore winds may blow normally pelagic species such as shearwaters, petrels or auks far inland. Some will die as a result, but many can reorient themselves and find their way back to the open ocean.

10 · HOW DO WE RELATE TO BIRDS?
Birds and people

BIRD NAMES
How did birds get their English names?

In many varied ways. Some birds were named for physical characteristics such as colour (Greenfinch), size (Great Tit) or markings (Whitethroat); some for their voice (Chiffchaff); some for what they do (Treecreeper); and others for where they live (Barn Owl). Ornithologists later developed new techniques for naming less common species. These included borrowing foreign names (such as Hobby, which comes from an Old French word meaning 'to jump'), adapting scientific names (such as phalarope, which derives from the Latin phalaropus, meaning 'coot-footed'), commemorating the place where the bird was first found (Dartford Warbler), or commemorating the person who discovered the species (Montagu's Harrier).

Treecreeper

What about more obscure names such as Wheatear or Redstart?

The origins of some other names are trickier to pin down. Many originated from Old or Middle English, and while they doubtless made perfect sense at the time, they sound pretty odd today. But a little linguistic detective work reveals that, for example, Redstart

'Fulmar' derives from the Old Norse for 'foul gull'

means 'red tail', while Wheatear derives from a phrase meaning 'white arse'. Likewise, Fulmar means 'foul gull', from the bird's habit of vomiting foul-smelling gunk over intruders, and its superficial resemblance to a gull.

Do bird names ever change?

Yes, often. Check out any old bird book and you will come across archaic names such as Golden-crested Wren (Goldcrest), Lesser Pettichaps (Chiffchaff) and Hedge Sparrow (Dunnock). Names evolve with the

Lots in a name

The current longest English name of any bird is, at least in terms of complexity, the King-of-Saxony Bird-of-paradise, which has six words and 26 letters. However, several other species have names with 30 or more letters, including Middendorff's Grasshopper Warbler, Abyssinian Yellow-rumped Seedeater and champion of them all, at 32 letters, Northern Long-tailed Glossy-starling. Several now obsolete names were even longer and more complex, including Mrs Forbes-Watson's Black-flycatcher (now Nimba Flycatcher), Ceylon Orange-breasted Blue-flycatcher (now Tickell's Blue-flycatcher), and the Himalayan Golden-backed Three-toed Woodpecker (now simply Himalayan Flameback) – an incredible 40 letters long.

Do Birds have Knees?

English language, so while some have remained constant over centuries, others have fallen into disuse. A good example of gradual change is the Robin, which was originally known as 'Redbreast', and later as 'Robin Redbreast', until eventually the second part of the name disappeared.

names that apply on both sides of the Atlantic, such as 'loon' (instead of 'diver') and 'jaeger' (instead of skua). But traditionalists are not convinced. Many proposed changes, such as 'Acorn Jay' and 'River Kingfisher', provoked such a storm of protest that they died a quiet death.

Have there been any recent changes to bird names?

Yes, because birds, like most other things, have acquired different names in different parts of the English-speaking world. In recent years, ornithologists have attempted to standardise these names in order to prevent confusion between species – for example, by adding prefixes such as Northern (Wheatear) and Barn (Swallow), or by creating uniform

Black-throated Diver or Arctic Loon? Two names for the same species

How did birds get their scientific names?

This was a lot more systematic. The pioneer who developed the process of giving each bird (and indeed every other organism) a scientific name was Carl von Linné (usually better known as Linnaeus, an 18th century Swedish botanist. Linnaeus invented the system known as 'binomial nomenclature' (see Chapter 2), which gave each species a unique combination of two names, the first indicating its genus and the second its species. For instance, while Brits and Americans may argue the merits

of Great Northern Diver and Common Loon, to scientists this bird will always be *Gavia immer*.

Do scientific names always reflect common names?

No. In fact, some make nonsense of each other. Take a look at gulls, for example. The Black-headed Gull has the scientific name *ridibundus*, which means 'laughing' – the English name of a North American gull. Meanwhile the Mediterranean Gull has the scientific name *melanocephalus*, which means 'black-headed'. Incidentally, although the Mediterranean Gull does have a black head, the Black-headed Gull only has a brown hood. And outside the breeding season neither has either.

Tongue-twisters

The longest scientific name of any bird is that of the Crowned Slaty Flycatcher from the Amazon Basin in South America – *Griseotyrannus aurantioatrocristatus*. In the UK and Europe, the longest scientific name is that of the Hawfinch – *Coccothraustes coccothraustes*, which means 'kernel-breaker' (describing its impressive nut cracking ability). But these pale into insignificance beside the longest scientific name of any organism, a Russian amphipod – *Brachyuropskkyodermatogammarus grievlinggwmnemnotus*.

There are several contenders for the shortest scientific name of any bird, each with only eight letters. These include *Crex crex* (Corncrake), *Tyto alba* (Barn Owl) and *Grus grus* (Common Crane). But all these are beaten by a mammal the Great Evening Bat of South-east Asia, whose scientific name is just four letters long: *Ia io*.

Common Cranes

How many birds are named after people?

According to Bo Beolens and Michael Watkins, authors of *Whose Bird?*, more than 2,500 different species or subspecies are named after people. The exact number is impossible to say, since the origins of some names remain shrouded in mystery. *Whose Bird?* covers 1,100 different individuals, many of whom – such as Alexander Wilson or Peter Simon Pallas – had several different species named after them. You probably won't be surprised to learn that the vast majority of these individuals are men, reflecting their dominance during the period when most species were given their names – the 18th and 19th centuries.

description. Unfortunately, very few species are being discovered today, so it is getting harder and harder to immortalise yourself in avian nomenclature.

Mrs Gould's Sunbird – named after John Gould's wife

Are any birds named after women?

A few, but these women were mostly the wives, sisters, mistresses or daughters of famous men, rather than well-known ornithologists in their own right.

How do you get a bird named after you?

Simple: just go out and discover it and then persuade a fellow scientist to name it in your honour. You could do this by 'collecting' a specimen of a new species in the field, or by spotting it among a pile of old museum skins. However you get your hands on it, just make sure that you write up the first formal scientific

Who has the most species named after them?

Twenty-six people – all men, of course – have at least 10 different species or subspecies named after them. Of these, one was Dutch, one Italian, two French, three American, four German, and a remarkable 15 were British. The Brits, as we might expect, occupy the top three places in the list. The bronze medal goes to P. L. Sclater, first editor of the *Ibis* (Britain's oldest ornithological journal), with 19 species. Charles Darwin, with 21 birds, claims a respectable silver. But the undisputed gold medallist, with a grand total of 24 different birds, is Victorian bird artist John Gould. He even managed to name one after his wife: Mrs Gould's Sunbird.

Were kites and cranes named after birds or vice versa?

As with flight itself, birds came first. So mechanical kites were named after their avian counterparts, whose twisting flight they emulated, while metal cranes got their name from the statuesque birds whose vertical posture they shared.

How many ornithologists gave their names to secret agents?

OK, so there's only one. But it's the big one! Author Ian Fleming was a neighbour and close friend of the author of *Birds of the West Indies*, a certain Mr James Bond. Legend has it that inspiration struck when Fleming glanced at his friend's book while searching for a name for 007. The rest is literary and cinematic history.

Professional fowl

The earliest bird domesticated by man was the Red Junglefowl of Asia, which was first domesticated in India more than 5,000 years ago. The Ancient Egyptians also domesticated many animals, including the Egyptian Goose, at least 4,000 years ago, and possibly as long as 4,500.

BIRDS AS FOOD AND PETS

What is a 'domesticated' species?

A domesticated species is one that has been bred in captivity over time in order to produce a product or products useful to people: usually its meat, eggs or feathers. This definition excludes birds kept purely for ornament, such as most cage birds, or for sport, such as hawks and falcons (though racing pigeons are usually considered to be domestic birds). In some cases, such as farmyard hens and geese, it is obvious that a bird is domesticated; in others, such as game birds or wildfowl bred and released for shooting, the boundary between domesticated and wild is not so clear. Typical domesticated species

Domestic chickens – the commonest bird in the world

in Britain and North America include Greylag and Swan Geese (from Europe and Asia), Mallard (from Europe, Asia and North America), Muscovy Duck (from South America) and various kinds of guinea fowl (from Africa). In some cases you can easily tell an individual's wild ancestry; in others, extensive selective breeding or hybridisation has made it all but impossible.

How many different species are domesticated?

Tricky question. Not only is domestication hard to define, but the variety of uses to which birds are put in different cultures also means that any list of domesticated species is likely to be incomplete. We can all agree on one though: the Red Junglefowl of Asia – today better known in deep-fried, casserole or McNugget form.

What is a 'game bird'?

Strictly speaking, it is any species unfortunate enough to find itself a target, for sport, food or profit, though the term is most commonly used for members of the order Galliformes (such as pheasants, grouse and guinea fowl). In Britain, the best-known game birds are Pheasant, Red Grouse and the two species of partridge, but others include Snipe and Woodcock, various species of duck and even Woodpigeon.

Apart from food, what else have we used birds for?

Throughout our history, human beings have exploited wild and domesticated birds in as many ways as possible. We have used feathers and skins for clothing (both ceremonial dress and fashion), oil from the body fat of seabirds for heating and cooking, and even bills and feet as primitive jewellery. Perhaps the most enterprising exploiters of birds were the islanders of St. Kilda, who until the 1920s lived almost entirely by harvesting seabirds, which they used for food, fuel and clothing – even wearing entire Gannet skins as slippers.

What about birds as pets?

Human beings have kept birds behind bars for many thousands of years – either for the beauty of their plumage or for their song. Our ancestors originally caught and kept common native songbirds such as the Linnet or

Pheasants are shot in their millions in Britain for 'sport'

Goldfinch. Then, as foreign travel and trade grew during the seventeenth and eighteenth centuries, tastes turned to more exotic species such as the Canary (from the Canary Islands), the Budgerigar (from Australia) and numerous small finches (mainly from Africa and Asia). Parrots, with their longevity and powers of speech, became a favourite companion for sailors – as in Long John Silver's famous sidekick in Robert Louis Stevenson's Treasure Island. Today a vast range of bird species, including many very rare ones, are kept in captivity.

Can all kinds of bird be kept in cages?

Almost any kind of bird can, theoretically at least, be kept in a cage. But keeping it alive is a different matter. For many specialised species, captivity would spell an early death. Some, such as swifts or albatrosses, spend most of their lives airborne, so a cage would be completely impractical – not to say fatal – for them. It is, of course, usually illegal to take birds from the wild and many rare species cannot be legally imported or bought and sold.

Does the cage bird trade do birds any harm?

Sadly, yes, if the birds are taken directly from the wild. Although many cage birds are now bred in captivity, there is still a great demand for wild birds,

Mynas are often kept in cages as they are able to learn human speech

especially species that are hard to breed. Unfortunately the most threatened species are often the prize targets of unscrupulous collectors, who will pay a fortune for illegally obtained specimens, such as rare Saker Falcons, which are in demand for hunting by falconers in several Middle Eastern countries. This fuels a flourishing underground trade, in which everyone from local collectors to international cartels has a vested interest in avoiding detection.

Which species are most affected?

BirdLife International estimates that roughly three out of every 10 globally threatened species are exploited by people in some way. Sometimes this is

for food, but mostly it is for the cage bird trade. Families hit hardest include parrots and macaws, pigeons and doves, and pheasants. The world's black spots include southeast Asia (especially China, Indonesia and the Philippines) and South America (especially Brazil).

Parrots are amongst the most popular cage birds

How much is the cage bird trade worth?

Today the cage bird trade is too huge to quantify. But to put it in perspective, some observers have estimated that the global trade in wildlife products as a whole, including live specimens, is second only to that in illegal drugs. CITES (the Convention on International Trade in Endangered Species) estimates that several million live birds of more than 4,000 different species are bought and sold every year.

The swarm

The most destructive pest species is undoubtedly the Red-billed Quelea, of the African savannah. The world's most numerous bird, flocks millions strong have been known to strip fields of crops bare, their destructiveness earning them comparison with a plague of locusts.

Have we ever worshipped birds?

Of course. We've worshipped most things in our time, so why not birds? Many early cultures reserved a sacred place for birds, or at least bird-like deities. Amongst these were the god Quetzalcoatl, the 'feathered serpent' worshipped by the Aztecs of central Mexico prior to the Spanish conquest, and various Egyptian gods, including Horus – often depicted with the body of a man and the head of a falcon.

Who first painted birds?

Well, we don't know the actual artist, but cave paintings in southern Europe that date back at least 18,000 years, depict recognisable images of owls and long-legged water birds (probably cranes, herons or storks).

Which is the first bird mentioned in The Bible?

A Raven, mentioned in the 8th chapter of the Book of Genesis: "And he sent forth a raven, which went forth to and fro, until the waters were dried up from off the earth." Noah seems to have kept no record of the actual species.

How many species of bird are mentioned in The Bible?

Somewhere between 25 and 40 different kinds of bird make it into the Old and New Testaments, though

A statue of Horus, the falcon-headed god, from Ancient Egypt

The Turtle Dove is just one of many species mentioned in the Bible

problems with translation and lack of precision (e.g. 'hawk', 'owl') mean that we cannot be sure exactly how many species this represents. The dove gets the most references – probably referring to the domesticated pigeon, although the Turtle Dove, a common migrant through the Holy Land, also crops up several times. Others include stork, pelican and eagle, and even the Ostrich – now extinct in the Middle East.

How many species are mentioned in Shakespeare?

At least 50 kinds of bird were immortalised by the Bard, although the same caveats apply as to the Bible. Shakespeare often compared people to birds: in *Much Ado About Nothing*, he writes that "Beatrice, like a lapwing, runs, close by the ground", while

Richard II has an eye "as bright as is the eagle's." Other bird references need a bit more deciphering, as in Hamlet's line "I know a hawk from a handsaw", in which the latter is a corruption of 'hernshaw', meaning a young heron.

Loadsamoney!

The world's most valuable bird was a racing pigeon called Invincible Spirit, bought by the British company Louella Pigeon World in 1992 for an extraordinary £110,000. Like ultra-valuable racehorses, it was immediately retired and put out to stud.

Which species are most popular in English poetry, and why?

The Skylark and the Nightingale both frequently crop up in the work of English poets – especially the Romantics, such as Shelley and Keats. They are used to evoke a particular mood: the Skylark generally one of joy and optimism, and the Nightingale one of mystery and melancholy.

Which English poet wrote most about birds?

Undoubtedly the nineteenth century poet John Clare, described by the ornithologist and broadcaster James Fisher as "the finest poet of Britain's minor naturalists as well as the finest naturalist of Britain's major poets". Clare observed at least 119, and possibly as many as 145 different species of bird in the area around his Northamptonshire home, and featured many of these in his poems.

The Kestrel used to be known as the 'windhover', the title of a poem by Hopkins

Being written without the aid of binoculars, his poems tend to focus on the behaviour of a bird rather than its plumage. Another poet inspired by birds was the Jesuit priest Gerard Manley Hopkins, whose poem *The Windhover* uses the perfection of a hovering Kestrel to evoke the wonders of creation.

The phrase 'bald as a coot' refers to the frontal shield on a Coot's head

Why are birds so popular in sayings and proverbs?

Probably because birds are among the most ubiquitous and visible aspects of the natural world, so people saw them more often than most other creatures, and observed their appearance and habits more closely. Thus sayings such as 'up with the lark', 'bald as a coot' and 'out for a duck' have become part of our day-to-day language, with few people stopping to consider their origins.

What about birds and weather folklore?

When a particular bird's appearance coincided with a regular event, such as geese arriving during a cold spell, or Swallows returning around the time of St George's Day, this soon passed into the calendar of folklore associated with the seasons. Much of this has survived to the present day, including the (sadly mistaken) belief that the early arrival of wintering wildfowl foretells a harsh winter to come.

Why are Robins associated with Christmas?

The early postmen of the Victorian era wore red uniforms, and were soon nicknamed 'robins'. Later on, Christmas cards were produced depicting a Robin delivering the mail; and the image stuck. It helps that Robins are particularly conspicuous around our gardens in winter.

How else do birds feature in popular culture?

How many ways do you need? Just think of Swan Vestas matches, Famous Grouse whisky, Kestrel lager or Penguin Books. Then there are cartoon characters such as Donald Duck and the Roadrunner ("beep beep!"), and hundreds – perhaps thousands – of pop songs featuring birds in their titles or lyrics: from 'My little chickadee' and 'Rockin' Robin' to Vera Lynn's 'Bluebirds over the White

Donald Duck – one of the best-known birds in the world!

Cliffs of Dover' (which were probably meant to be Swallows). Not to mention the various groups named after birds: including the Guillemots, Doves and of course the Eagles.

Why are girls sometimes referred to as 'birds'?

According to the Oxford English Dictionary, the slang use of 'bird' to mean 'girl' dates back to the fourteenth century, though its modern usage did not really get under way until the 1960s, when it was common to hear young men refer to their girlfriend as 'me bird'. Another piece of '60s slang for a girl was 'chick', which survives today in expressions such as 'chick-lit' or 'chick-flick'. I'll leave sociologists to ponder the reasons for this.

PROTECTING AND STUDYING BIRDS

Which was the first bird to be protected by law?

Probably the Sacred Ibis. The Ancient Egyptians worshipped many gods, including Thoth, who was often represented as an ibis. Ibises would appear at the time the River Nile rose, and were therefore seen as bringing good fortune, preserving the country from plagues and serpents. Mummified remains of ibises have been found in tombs at Thebes and Memphis. Because of this association with the gods, it was considered a terrible crime to kill an ibis, punishable by death.

A Brown Pelican, one of the waterbirds protected in the USA's first wildlife refuge

The Sacred Ibis was worshipped by the Ancient Egyptians

When was the first bird reserve established?

In 1903, President Teddy Roosevelt designated Pelican Island, on Florida's east coast, the USA's first National Wildlife Refuge, primarily to protect nesting waterbirds. In Britain the RSPB's first bird reserve was founded on Romney Marsh in 1930, with that at nearby Dungeness following soon after in 1932. Cley Marshes in Norfolk had been protected since 1926, but this was originally for shooting purposes rather than to benefit the birds themselves.

When was the RSPB founded, and why?

The RSPB began in protest against the use of egret feathers in ladies' fashions

The RSPB (Royal Society for the Protection of Birds) began in a Manchester suburb in February 1889. Its founders, who each paid a subscription of two old pence, were a small group of women, horrified at the wearing of bird plumes and skins from species such as egrets as fashion accessories by their peers. By 1899 the Society had over 150 branches and more than 20,000 members; today it has over one million.

Why do we ring birds?

Ringing (known in the USA as banding) is the practice of fitting a small metal ring around a bird's leg, in order to learn more about its behaviour, movements, life cycles and – especially – migrations. Data from ringed birds (either recoveries when they are dead or those retrapped while still alive) help us to build up a picture of many different aspects of an individual bird's life, and – taken together with other recoveries – enable us to draw conclusions about the species' general behaviour. Ringing can tell us how long birds live, where they spend the winter, how fast they travel when migrating, and many other things of great importance to biologists and conservationists.

Ringing birds provides crucial information about their journeys and lifestyles

Does ringing harm the bird?

Hardly ever. Very occasionally a captured bird may be so exhausted that it suffers trauma during handling. And, as in any interaction between humans and wild creatures, accidents can happen. But these are extremely rare, and in general ringing does no harm to the birds being caught. Certainly the benefits that ringing brings to bird conservation far outweigh the risk of harming individual birds. Incidentally, ringers are licensed and to become one takes years of rigorous training.

When did ringing begin?

The first person to mark birds using 'rings' around their legs was the German ornithologist Johann Leonard Frisch, who in 1740 tied coloured threads to the feet of Swallows. When

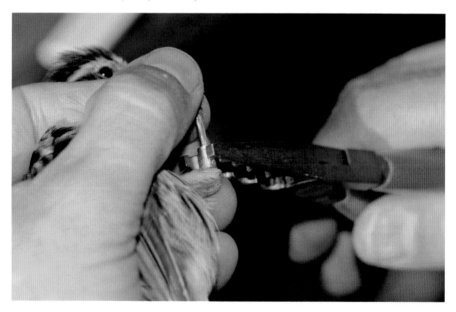

A Black-headed Gull wearing a ring

the colour in the threads did not run, Frisch proved that the birds did not hibernate under water, as had been thought. During the early 19th century John James Audubon tied a silver wire around the leg of an Eastern Phoebe, but did not follow up on his experiment. Then, in 1890, a private landowner in Northumberland used aluminium rings for the first time, placing them round the legs of young Woodcocks in order to study their movements.

How did this develop into a formal scheme?

Early in the 20th century, various ornithologists in both Britain and North America began official schemes to ring birds. Today, the British scheme is run by the British Trust for Ornithology (BTO), while the North American Bird Banding Program is jointly administered by the US Department of the Interior and the Canadian Wildlife Service. The rings used are generally made from aluminium alloy, making them light but strong.

How many birds have been ringed in Britain and North America?

From the first official British scheme in the early 20th century, to the end of 2014, more than 42 million birds, of well over 400 different species, have been ringed. (In 2014 alone over one million birds of 253 species were ringed). But ringers across the pond can better this: to date, well over 60 million individual birds have been banded in North America, of which more than three million have been recovered – almost five per cent of the total (and representing more than twice the success rate of the Brits).

Which species is ringed most frequently?

Blue Tits, by a long way. In Britain and Ireland this species almost always has more individuals ringed in a year than any other. In 2014, for example, 120,343 Blue Tits were ringed, almost twice as many as its nearest challengers, Great Tit (68,862) and Blackcap (58,313).

The Blue Tit is by far the most frequently ringed bird in Britain

How many ringed birds are recovered?

Not many. At least, not as a percentage of the total ringed. Of all the birds ever ringed in Britain, approximately 935,000 have been 'recovered' (either found dead, or retrapped), representing about 2.2 per cent of the total. Of course, the figures vary from year to year and species to species. The proportion recovered is always much higher for larger birds such as ducks and waders, or colonial nesters such as storm-petrels and Sand Martins, but drops to well below one in 100 for smaller migrants such as warblers.

What is 'colour ringing' for?

Ornithologists sometimes put brightly coloured rings on birds in order to distinguish between individuals in a particular population, so that their behaviour can be studied more easily. In the case of larger birds such as gulls or waders, coloured dyes are often used, while large raptors are sometimes marked using plastic wing-tags. From the late 20th century onwards larger birds such as eagles, storks and cranes were also 'radio-tagged', giving us invaluable data on their movements and migrations. Today, thanks to the miniaturisation of technology, GPS trackers can be put on birds as small as a Nightingale, enabling us to gain a remarkable insight into their exact route and the timing of their journeys.

What should I do if I find a ringed bird?

If you find a ringed bird, you should report it as soon as possible to the British Trust for Ornithology. Either call 01842 750050, or log on to the BTO's website: www.bto.org/ringing and follow the instructions.

What is a bird observatory?

A place where birders and ornithologists go to observe and record birds – especially during migration periods. Observatories usually comprise a building and a trapping area, and are commonly situated on a coastal headland or island in order to maximise the number of migrants and different species likely to show up. Well-known observatories include Fair Isle and Dungeness in Britain, Falsterbo in southern Sweden, and Cape May in New Jersey, USA. The first bird observatory was set up by Heinrich Gatkë on the German island of Heligoland in the 19th century.

The Florida Scrub Jay has been studied longer than almost any other bird

What is 'ethology'?

Ethology is the study of animal behaviour, developed between the First and Second World Wars by a new generation of 'observer-scientists', who studied wildlife (notably birds) in the field rather than in the museum or laboratory. Pioneers included Sir Julian Huxley, who spent his summer holidays in 1916 watching the breeding behaviour of Great Crested Grebes; Niko Tinbergen, a Dutch scientist who concentrated mainly on gull colonies; and Konrad Lorenz, an Austrian who popularised the new science through best-selling books such as *King Solomon's Ring*, published in 1952. At first, ethologists were looked down on by their scientific peers. But gradually the new science earned respect, culminating in the award of the Nobel Prize for Physiology or Medicine to Lorenz, Tinbergen and Karl von Frisch in 1973.

Which is the most intensively studied bird?

In Britain and Europe, the Great Tit seems to have proved endlessly fascinating. It has been studied in

Wytham Wood in Oxfordshire since 1947. In North America, it is probably the Florida Scrub Jay, studied in central Florida since the 1950s. Incidentally this bird is the only species entirely restricted to the state of Florida.

What is 'oology'?

Oology is the 'pseudo-science' of egg-collecting. For much of the 19th and early 20th centuries this was a respectable and popular activity, very much part of the ornithological establishment. However, as bird protection gradually gained in popularity, egg-collecting declined, and was finally made illegal under various Bird Protection Acts following the Second World War. It is now considered completely unacceptable, although birds' eggs are still studied by scientists, for example in order to detect the effects of pesticides on reproduction.

225

WATCHING BIRDS

Who was the first birdwatcher?

There are several candidates for this honour, depending on what we mean by 'watch'. As well as the anonymous cave-painters of roughly 18,000 years ago, and Noah, who sent forth a raven and dove from the Ark, there is the scientist and philosopher Aristotle (384–322 BC), who made many accurate (and a few inaccurate) observations of birds. If we regard birdwatching as being the observation of birds primarily for pleasure, as distinct from study, then I would suggest the Reverend Gilbert White, Vicar of Selborne in Hampshire. White lived from 1720 to 1793, and wrote one of the best-known natural history books in the English language, *The Natural History of Selborne*, a diary of nature observations from his parish.

What was the first bird book in English?

The first bird book written entirely in English was *The Ornithology of Francis Willughby*, published posthumously in 1678 and edited by Willughby's friend John Ray.

When did the term 'birdwatching' first appear in print?

As the title of the book *Bird Watching*, by Edmund Selous, published in 1901.

What is 'fieldcraft'?

It is simply an umbrella term for a range of skills and strategies used by experienced naturalists in order to get closer to birds and other wildlife.

Many reserves now provide boardwalks from which people can enjoy the birds

It encompasses both knowledge and behaviour: from knowing what time of day you are most likely to see a particular species, to avoiding sudden movements when approaching a roosting flock. It includes many tricks of the trade, gained by years of experience in the field.

A hide at the RSPB's Conwy reserve, in North Wales

Who wrote the first field guide?

During the 19th and early 20th centuries, several portable bird books were published in Britain and North America. But the first proper field guide is generally held to be the aptly-named *A Field Guide to the Birds*, written and illustrated by Roger Tory Peterson, which covered the birds of eastern North America, and was published in 1934. Peterson's technique was simple but effective: his illustrations used arrows to indicate 'field marks' (the key identification pointers for each species), and depicted each bird in a standardised pose for easy comparison between species. The guide revolutionised birdwatching in North America, selling more than three million copies since publication.

Which was the first British and European field guide?

The first field guide to cover all regularly occurring birds in Europe was *A Field Guide to the Birds of Britain and Europe*, by Peterson, Mountfort and Hollom, which appeared in 1954. However, it was beaten to the title of first British field guide by the *Pocket Guide to British Birds*, by Fitter and Richardson, which was published in 1952.

What is 'jizz'?

Jizz is a term used by birdwatchers that refers to a bird's general impression and behaviour, rather than any specific feature, and can be used to identify a bird quickly at a brief glimpse or a great distance. Jizz is difficult to pin down; it amounts to a combination of the observer's experience with some indefinable but definite quality on the part of the bird – essentially its 'character'. The origin of the term has been disputed, but it appears to have been coined by the Manchester ornithologist T.A. Coward, who wrote about jizz in 1922.

What is 'pishing'?

Pishing is a technique that birders use in order to attract small birds such as warblers and tits. It can be done in several ways, either by kissing the back of your hand (also called 'squeaking'), or by making a repeated 'pish-pish-pish' sound. Results are not guaranteed, though curiously it generally seems to work better in North America than in Britain. The noise supposedly provokes the curiosity of small birds, especially those which habitually travel in flocks – perhaps by resembling an alarm call, which stimulates them to gather and mob a predator. When it fails, however, pishing can seem pointless and faintly ludicrous.

What is 'listing'?

The keeping of lists of birds seen either over a period of time (e.g. a day, year or life), or in a particular place (e.g.

Most birders carry a field guide in order to identify the birds they see

Birdwatching - or birding as it is often known - is one of the most popular of all leisure activities

garden/backyard, local patch, county, country, continent or the whole world). Often it is some combination of the two (e.g. British list for 2015; September list for Kent). Lists can be tailor-made for every occasion. One popular favourite is the New Year's Day list – the original of which was created by H.G. Alexander on 1st January 1905, when he saw a grand total of 17 species.

What is 'twitching'?

Twitching is a slang term for the dedicated pursuit of individual rare birds, usually those which have been blown off-course and turned up somewhere far from home. It is inextricably linked to listing, since twitching makes the list grow longer. Thus British twitchers might travel miles to see a North American Laughing Gull in Britain because they need it for their British list, despite the fact that it is a common bird in much of the United States. Twitchers are driven by a combination of factors, including one-upmanship, an anal obsession with lists, the companionship of a tribe, or simply the thrill of the chase. They usually do little harm, apart from the occasional trespassing incident, and are often expert at identifying birds – especially rare ones.

What do 'dip', 'grip' and 'string' mean?

British twitchers have even evolved their own slang. To 'dip' or 'dip out' is to miss a rare bird, either by not hearing about it in the first place, or worse still, by travelling a long distance and arriving after the bird has left. To 'grip someone off' has the opposite meaning: you have seen the bird but your friend or rival has not. Finally, 'stringing' is the cardinal sin: to either pretend to have seen a rare bird, or to delude yourself that you have done so. A serial stringer soon becomes persona non grata in the twitching world.

How many people watch birds?

In Britain, more than one million people (about two per cent of the population) are members of the RSPB, though perhaps only a minority of these could be described as active or regular birdwatchers. In addition, many more people regularly feed birds in their gardens. In the US, the number of regular birders has been estimated at anywhere between 300,000 and 1.3 million; though one survey claimed that there were as many as 46 million – about one in six of the total population. It's almost impossible to define what constitutes a birdwatcher. But, judging from the sheer number of bird books alone, few other creatures receive more scrutiny than birds.

Why do we watch birds?

I can only quote two people, one from each side of the Atlantic, who have thought about the answer longer and more deeply than I have. In 1940, ornithologist James Fisher wrote that "The observation of birds may be a superstition, a tradition, an art, a science, a pleasure, a hobby, or a bore; this depends entirely on the nature of the observer." Fifteen years earlier, in 1925, the American writer Donald Culross Peattie had a more poetic view of why we are so obsessed with these wonderful creatures: "Man feels himself an infinity above those creatures who stand, zoologically, only one step below him, but every human being looks up to the birds. They seem to us like emissaries of another world which exists about us and above us, but into which, earth-bound, we cannot penetrate." I can think of no better sentiment with which to end this book.

Epilogue: Do birds have knees?

Yes, but not where some people think they are! A bird's knee is actually concealed in the plumage at the top of its leg, whereas the joint that is often assumed to be the knee (roughly halfway up the visible leg on long-legged birds) is in fact the anklebone. This has led to the popular misconception that birds' knees bend backwards…

Opposite: The Spotted Dikkop – also known as a 'thick-knee' – but where are its knees?!

BIBLIOGRAPHY

Beolens, B. and Watkins, M. 2003. *Whose Bird?* Christopher Helm, London.

Bird, D. M. 2004. *The Bird Almanac.* Firefly Books, Buffalo, New York.

Birdlife International. 2000. *Threatened Birds of the World.* Lynx Edicions and Birdlife International, Barcelona and Cambridge.

Brooke, M. and Birkhead, T. 1991. *The Cambridge Encyclopedia of Ornithology.* Cambridge University Press, Cambridge.

Campbell, B. and Lack, E. 1985. *A Dictionary of Birds.* Poyser, Calton.

Catchpole, C.K. and Slater, P.J.B. 1995. *Bird Song: Biological themes and variations.* Cambridge University Press, Cambridge.

Clements, J. 2000. *Birds of the World: a Checklist: Fifth Edition.* Ibis Publishing, Vista, California.

Gibbons, D.W., Reid, J.B. and Chapman, R.A. 1993. *The New Atlas of Breeding Birds in Britain and Ireland.* Poyser, London.

Leahy, C. 2004. *The Birdwatcher's Companion to North American Birdlife.* Princeton University Press, Princeton and Oxford.

Lockwood, W.B. 1984. *The Oxford Book of British Bird Names.* Oxford University Press, Oxford.

Mearns, B. and Mearns, R. 1988. *Biographies for Birdwatchers.* Academic Press, London.

Richards, A. 1980. *The Birdwatcher's A–Z.* David & Charles, Newton Abbot.

Snetsinger, P. 2003. *Birding on Borrowed Time.* ABA, Colorado Springs.

Skutch, A. 1996. *The Minds of Birds.* Texas A&M University Press, College Station.

Todd, F. 1994. *10,001 Titillating Tidbits of Avian Trivia.* Ibis Publishing, Vista, California.

Weaver, P. 1981. *The Birdwatcher's Dictionary.* Poyser, Calton.

ACKNOWLEDGEMENTS

First, I'd like to thank my great friend Nigel Redman, who came up with the idea of this book in the first place. Also at Bloomsbury, Julie Bailey, who commissioned the book, and Jane Lawes, who oversaw the manuscript to its conclusion. Mike Unwin edited the original text, adding clarity and structure. Thanks also to Rod Teasdale, who designed the book, and Marianne Taylor for proofreading.

PHOTO CREDITS

INDEX

A
abmigration 205
accidentals 203
adaptive radiation 56
Aepyornis (Elephant Bird) 19, 104, 176
Albatross, Black-browed 204
 Royal 31, 41, 182
 Wandering 31, 177, 182, 187
albinism 22
anatomy 6–13
Antpitta, Alta Floresta 58
Archaeopteryx 43
Astrapia, Ribbon-tailed 18
Auk, Great 103
Avocet 121

B
Bananaquit 49
Bat, Pipistrelle 55
bathing 17
beaks 9
Bee-eater 203
 Little Green 90
bellbirds 143
Bible 216–17
bills 9–12
 feeding 121
binocular vision 28, 30
binomial nomenclature 51–2, 208–9
Biological Species Concept 52, 53–4
bird topography 16
Bird-of-paradise, Greater 93
 King of Saxony's 207
birds 6
 best dawn chorus 141
 best mimic 151
 champion bird-feeding nation 134
 child-eating 128
 continent with highest number of species 76
 countries with highest number of species 75
 deepest diving 107
 earliest and latest breeding ages 157
 earliest domesticated birds 211
 fastest flying bird 101
 fastest running 109
 fastest swimming 106
 fewest number of species per country 79
 greatest wingspan 31
 highest and lowest altitude 90
 highest flying bird 103
largest clutch of eggs 180
largest eggs 176
largest ever bird 19
largest family of birds 49
largest flying bird 26
largest food item 130
largest living bird 8
largest nests 169
largest number of subspecies 56
largest order of birds 45
largest prey items 124
longest bill 11
longest English names 207
longest fledging period 187
longest incubation period 182
longest legs 14
longest migratory journeys 193
longest penis 165
longest scientific names 209
longest tails 18
longest tongue 12
longest-lived birds 41
loudest song 143
most airborne bird 95
most carrying sound 147
most eggs per season 185
most endemic species per country 93
most food eaten relative to body weight 119
most northerly birds 83
most number of rare species per country 79
most numerous bird ever 69
most numerous raptor 65
most numerous seabird 66
most numerous wild bird 62
most songs in a single day 144
most southerly birds 84
most successful introduced species 93
most valuable bird 217
most varied repertoire of song 137
most widespread wild bird 88
rarest bird 73
shortest migratory journeys 200
smallest birds 33
smallest repertoire of song 139
Bittern, Eurasian 147, 152
Blackbird 93, 137, 138, 143, 154
Blackcap 122, 149, 201–2, 224
Blue-flycatcher, Tickell's 207
Bond, James 211
Booby, Masked 130, 177

Red-footed 130
Boubou, Tropical 150
breathing 8
breeding 154–7
 courtship displays 162–3
 hybridisation 166–7
 mating 157, 166
 pairing 157, 160–5
 territories 157–60
Bristlehead 49
brooding 183
Bunting, Corn 141, 164, 187
 Indigo 164
 Snow 77, 83
Bustard, African Kori 26
 Great 26
Buzzard 98, 123, 174
 Honey-buzzard 124

C
cage birds 213–15
calls 137, 149–52
 echolocation 153
 quacking 153
camouflage 21
Canary 145, 213
Capercaillie 90, 136
Catbird, Gray 145
cats 67
Chaffinch 30, 146, 148, 190
Chickadee, Mountain 77
Chiffchaff 137, 140, 149, 153, 206, 207
 Iberian 148
Chough 153
 Alpine (Yellow-billed) 90
cisticolas 153
classification of birds 44, 46–8
 binomial nomenclature 51–2, 208–9
 Biological Species Concept 52, 53–4
 families 48
 genus (genera) 48
 lumping and splitting 52
 orders 45
 Phylogenetic Species Concept 52, 54–5
 species 50–1, 55–6
climate change 194
climbing 109
clines 87
Cobra, King 168
Cock-of-the-rock, Guianan 162
Cockatoo, Sulphur-crested 41
colonial nesting 173, 186
colour 20–3

can birds see colour? 29
commensalism 127
communication 136
 calls 137, 149–52
 flocking 112
 non-vocal sounds 152
 song 136–49
Condor, Andean 26, 31
 California 73
convergent evolution 57
Corncrake 57, 209
courtship displays 162–3
 lekking 165–6
Cowbird, Brown-headed 185
Coypu 92
Crane, Common 209
 Whooping 73
Crossbill, Common/Red 50, 155
Crow, Carrion 174
 Indian House 92
 Jungle/Japanese 38
 New Caledonian 38
Cuckoo-Roller 49
Cuckoo, Common/Eurasian 153, 182, 194, 199
culture and birds 218–19
Curlew, Slender-billed 72

D
dawn chorus 141
digestive system 119
Dikkop, Water 174
dinosaurs 42
dippers 107
dipping 230
diseases 36
distribution 74–8
 changes in range 82–5
 endemic species 79–83
 indigenous and introduced species 87–93
 species variation 86–7
Diver, Great Northern 209
diving 107
Dodo 103
domesticated birds 211–13
 cage birds 213–15
Dotterel 77, 161
Dove, Collared 82
 Turtle 194, 217
drinking 131–2
Duck, Black-headed 182
 Blue-billed/Lake 165
 Long-tailed 153, 174
 Mandarin 89

Do Birds have Knees?

Muscovy 212
Ruddy 89, 92
Whistling 153
White-headed 92
Wood 89
Dunlin 121, 174
Dunnock 164, 182, 207
dust-bathing 17

E
Eagle, Bald 16, 129, 159, 169, 170
Crowned 124, 128
Golden 128, 159, 180–1
Harpy 124
Short-toed 85
Tawny 122
White-tailed 90, 129
ears 30
echolocation 153
eclipse plumage 25
eggs 175–9
abandonment 179
brood parasitism 181–2
brooding 183
hatching 180–1, 183
incubation 179–82
Egret, Cattle 78, 82, 88, 127
Eider, Common 101, 153
Elephant Bird (*Aepyornis*) 19, 104, 176
emotions 40
Emu 49
English literature 217–18
ethology 225
evolution 42–3, 57–8
excretion 120
extinctions 68–73

F
Falcon, Eleonora's 155
Red-footed 112
Saker 214
Falconet, Black-thighed 33
Collared 33
Philippine 33
White-fronted 33
falls 205
faunal areas 74–5
feathers 6, 15–17
feeding 116–17
bill shapes 121
diets 122–3
digesting food 119
finding food 118, 120

garden birds 133–5
hiding and storing food 126–7
predatory birds 123–5
using other animals 127–30
feet 13
predatory birds 125
webbed feet 14
why don't waterbirds get frostbite? 32
field guides 227
fieldcraft 226–7
Finch, Trumpeter 153
'Vampire' 130
Woodpecker 38
Flameback, Himalayan 207
fledging 184–7
flight 94–7
atmospheric conditions 100–1
how do young birds learn to fly? 187
techniques 98–9
V formation 196–7
flightlessness 102–4
flocking 110–13
Flowerpecker, Wakatobi 58, 59
Flycatcher, Crowned Slaty 209
Nimba 207
Spotted 194
fossil record 43
Fox, Arctic 174
Fulmar, Northern 41, 83, 207

G
Galapagos finches 38, 56, 58, 130, 167
game birds 213
Gannet 159, 173, 213
garden birds 133–5
global warming 67
Glossy-starling, Northern Long-tailed 207
Goldcrest 188, 207
Goldfinch 93, 213
Goose, Bar-headed 103
Bean 52
Brent 52
Canada 92, 167
Chinese 212
Egyptian 174, 211
Greylag 212
Pink-footed 52
Snow 205
White-fronted 205
Grebe, Great Crested 162, 225
Greenfinch 206
gripping 230
Ground-finch, Sharp-beaked 130

236

Grouse, Red 213
Guillemot, Brünnich's 197
    Common 22, 159, 178, 197
Gull, Black-headed 26, 209
    Herring 37
    Ivory 83
    Kelp 130
    Laughing 229
    Little 196
    Mediterranean 209
    Ross's 200

H
Hamerkop 49, 174
Harrier, Montagu's 206
hatching 180–1, 183
Hawfinch 209
Hawk-owl, Cebu 58, 59
Hawk, Red-tailed 84
head turning 28
hearing 30
heart rates 9
Heron, Green 38–9
hibernation 35
Hoatzin 45, 48, 49, 123
Hobby, Eurasian 125
Holarctic 75
Honey-buzzard 124
honeyguides 128
Hoopoe 203
Hummingbird, Bee 33, 176
    Sword-billed 11
hummingbirds 119
hybridisation 166–7
Hypocolius 49

I
Ibis, Sacred 220
Ibisbill 49
immature plumage 26
incubation 179–82
intelligence 37–40

J
Jaeger, Parasitic 22, 52
Jay, Florida Scrub 225
jizz 228
Junglefowl, Red 211, 212
juvenile plumage 26

K
Kagu 49, 104
Kakapo 147

Kestrel 84, 98, 218
Kingfisher 44
Kite, Black 65
    Red 112
    Snail 122
Kittiwake, Black-legged 83, 153
Kiwi, Brown 179, 182
    Little Spotted 176
kleptoparasitism 129
knees 230
Knot, Red 191, 196

L
Lammergeier 39
Lapwing 174
Lark, Horned (Shore) 88
    Razo 79
lekking 165–6
leucism 22
lifespan 35
Limpkin 49
Linnet 213
listing 228–9
Loon, Common 209
Lyrebird, Superb 143, 151

M
Macaw, Spix's 73
Magpie 66, 171
    Azure-winged 78
Mallard 153, 212
Manakin, White-bearded 152
Martin, House 171, 174
    Sand 194
mating 157, 166
melanism 22
Merganser, Red-breasted 101
migration 188–94
    abmigration 205
    finding their way 195–202
    flocking 111
    getting lost 202–5
    reverse migration 204–5
milk bottle raiding 39–40
mimicry 150–1
Mink 89, 92
Moa 19, 104
mobbing 114
Mockingbird, Northern 137
Moorhen 57
mortality 35–6
moulting 24–6
moving on land 108–9

Murre, Common 197
  Thick-billed 197

N
names 206–8
  bird sounds 153
  kites and cranes 211
  named after people 210
  scientific names 208–9
Needletail, White-throated 101, 193
nestlings 184–5
  baby birds fallen out of nest 186
  water 132
nests 168–9
  colonial nesting 173, 186
  different designs 170–1
  keeping clean 186–7
  neighbours and squatters 174
  nest building 171–2
  site fidelity 172
Nightingale 137, 140, 142–3, 146, 157, 218, 224
nocturnal birds 29, 115
numbers of birds 60–2
Nutcracker, Clark's 77, 200
nuthatches 109

O
observatories 224–5
Oilbird 49
Oilbird, South American 153
oology 225
Osprey 13, 44, 49, 78, 124, 129, 192, 196
Ostrich 8, 9, 13, 14, 19, 39, 45, 49, 102, 104, 109, 176, 178, 180
Ovenbird 139
Owl, Barn 29, 78, 174, 206, 209
  Burrowing 13
  European Eagle 123
  Long-eared 30, 123
  Screech 22, 153
  Snowy 174
  Tawny 29
owls 28, 29

P
pairing 157, 160–5
Palmchat 49
Parakeet, Monk 92
  Rose-ringed/Ring-necked 92
parasites 19
parasitism 130
  brood parasitism 181–2
Parrot, African Grey 151

Partridge, Grey 87, 180
  Red-legged 87
passage migrants 196
passerines 46
Peafowl 157
Pelican, Australian White 11
pelicans 121
Penguin, Emperor 84, 107, 168, 176, 182
  Gentoo 106
  King 107, 187
Peregrine 30, 78, 84, 101, 111, 114, 120, 174, 187
perspiring 8
Petrel, Antarctic 84
  Southern Giant 130
  Zino's 72
Pheasant 87, 91, 213
  Crested Argus 18
  Golden 91
  Lady Amherst's 91
Phoebe, Eastern 223
Phylogenetic Species Concept 52, 54–5
Pigeon, Feral 89
  Passenger 69
  Tooth-billed 9
pigeons, racing 217
Piping-guan, Trinidad 80
Pipit, Meadow 146
  Richard's 204
  Water 200
pishing 228
Plains-wanderer 49
Platypus, Duck-billed 175
Plover, Crab 49
  Magellanic 49
  Ringed 121
plumage 15–19
  moulting 24–6
  phases 22–3
  waterproofing 18, 106
Po'ouli 73
Poorwill, Common 35
populations 60–2
  changes to populations 63–7
problem-solving 38
protecting birds 220–1
proverbs 218
Ptarmigan, Rock 21, 77
Pygmy-Tyrant, Black-capped 33
  Short-tailed 33

Q
quacking 153
Quail, Bobwhite 180